Wireless Networks

Series Editor

Xuemin Sherman Shen, University of Waterloo, Waterloo, ON, Canada

The purpose of Springer's Wireless Networks book series is to establish the state of the art and set the course for future research and development in wireless communication networks. The scope of this series includes not only all aspects of wireless networks (including cellular networks, WiFi, sensor networks, and vehicular networks), but related areas such as cloud computing and big data. The series serves as a central source of references for wireless networks research and development. It aims to publish thorough and cohesive overviews on specific topics in wireless networks, as well as works that are larger in scope than survey articles and that contain more detailed background information. The series also provides coverage of advanced and timely topics worthy of monographs, contributed volumes, textbooks and handbooks.

** Indexing: Wireless Networks is indexed in EBSCO databases and DPLB **

More information about this series at https://link.springer.com/bookseries/14180

Qiang Ye • Weihua Zhuang

Intelligent Resource Management for Network Slicing in 5G and Beyond

 Springer

Qiang Ye
Department of Computer Science
Memorial University of Newfoundland
St. John's, NL, Canada

Weihua Zhuang
Department of Electrical and Computer
Engineering
University of Waterloo
Waterloo, ON, Canada

ISSN 2366-1186   ISSN 2366-1445   (electronic)
Wireless Networks
ISBN 978-3-030-88668-4   ISBN 978-3-030-88666-0   (eBook)
https://doi.org/10.1007/978-3-030-88666-0

This Springer imprint is published by the registered company Springer Nature Switzerland AG
The registered company address is: Gewerbestrasse 11, 6330 Cham, Switzerland

Preface

This book, entitled "Intelligent Resource Management for Network Slicing in 5G and Beyond," provides a timely and comprehensive study of developing efficient network slicing frameworks in both 5G wireless and core networks from various protocol stack layer perspectives, which includes virtual network topology design, end-to-end delay modeling, dynamic resource slicing, and link-layer and transport-layer protocol customization. Optimization and queueing analysis techniques are applied to solving different problems for network slicing. This book includes six rigorously refereed chapters, and the material serves as a useful reference for professionals from both academia and industry, and graduate students working on the area of network slicing in 5G (and beyond) networks.

In Chap. 1, we give an introduction to networking architectures of both 5G wireless and core to support diversified services in the Internet-of-Things era and then introduce the resource management in 5G from the perspectives of different protocol stack layers.

In Chap. 2, we study the topology design problem for each sliced virtual network and establish an end-to-end delay analytical model for each embedded service function chain.

In Chap. 3, we investigate the resource slicing problems for 5G core and wireless networks, respectively, to improve the resource utilization with quality-of-service isolation.

In Chap. 4, we design customized transport-layer protocols for time-critical and data-hungry applications, respectively.

In Chap. 5, we present adaptive and service-oriented medium access control protocols for Internet-of-Things-enabled mobile networks with the *ad hoc* networking mode.

In Chap. 6, we conclude this book with a summary and discuss future research directions in applying machine learning techniques to resource slicing and protocol automation for beyond 5G networks.

We would like to thank Prof. Xuemin (Sherman) Shen, Dr. Omar Alhussein, Dr. Jiayin Chen, Si Yan, Dr. Kaige Qu, Dr. Phu Thinh Do, Dr. Wei Quan, Dr. Junling Li, Dr. Weisen Shi, Dr. Peng Yang, Dr. Shan Zhang, Dr. Ning Zhang, Dr. Xu Li, Dr. Jaya Rao, Dr. Li Li, Dr. Philip Vigneron, Jayden Luo, Yushi (Alfred) Cao, Wenjing Chen, Andre Daniel Cayen, and Zichuan Wei for their helpful discussions and contributions to the research presented in this book, without which this book could have not been completed.

Waterloo, ON, Canada   Qiang Ye
August 2021   Weihua Zhuang

Contents

Acronyms

3GPP	3rd Generation partnership project
5G	Fifth generation
ACK	Acknowledgment
AP	Access point
AR	Augmented reality
ARIMA	Autoregressive integrated moving average
AV	Autonomous driving vehicle
BBUs	Baseband processing units
BIB	Buffer occupancy indication bit
BS	Base station
CapEx	Capital expenditure
CC	Caching-caching
CFP	Contention-free period
CN	Caching notification
CP	Contention period
CR	Caching-retransmission
C-RAN	Cloud radio access network
cRRC	Central radio resource control
CSMA/CA	Carrier sense multiple access with collision avoidance
CTP	Control period
D2D	Device-to-device
DCF	Distributed coordination function
DIFS	Distributed interframe space
DNN	Deep neural network
DNS	Domain name system
DRF	Dominant resource fairness
DR-GPS	Dominant-resource generalized processor sharing
DRL	Deep reinforcement learning
dRRM	Distributed radio resource management
D-TDMA	Dynamic time division multiple access
E2E	End-to-end

eMBB	Enhanced mobile broadband
GPS	Generalized processor sharing
HD	High definition
HOL	Head-of-line
IDS	Intrusion detection system
InP	Infrastructure provider
IoT	Internet of Things
LTE	Long-term evolution
M2M	Machine-to-machine
MAC	Medium access control
MANET	Mobile ad hoc network
MBS	Macro-cell BS
MILP	Mixed-integer linear program
MINLP	Mixed-integer nonlinear program
mMTC	Massive machine-type communication
MTC	Machine-type communication
MTD	Machine-type device
MU	Mobile user
NFV	Network function virtualization
OpEx	Operational expenditure
PDF	Probability density function
QoS	Quality-of-service
RD	Retransmission data
RL	Reinforcement learning
RR	Retransmission request
RRHs	Required retransmission hops
RTO	Retransmission timeout
RTT	Round-trip time
SBS	Small-cell BS
SDN	Software-defined networking
SFC	Service function chain
SINR	Signal-to-interference-plus-noise ratio
SNR	Signal-to-noise ratio
SVC	Scalable video coding
TCP	Transmission control protocol
UDP	User datagram protocol
URLLC	Ultra-reliability and low-latency communications
V2V	Vehicle-to-vehicle
VNF	Virtual network function
VoD	Video-on-demand
VR	Virtual reality

Chapter 1
Introduction

1.1 5G Networks

1.1.1 Internet-of-Things Era

With the advancement of mobile communication networks, the current cellular networks have been evolving to the fifth generation (5G). Different from the previous generations which mainly support the communications among mobile users (e.g., mobile phones, tablets, and vehicles), the 5G networks are developed to interconnect not only user terminals with enhanced mobility, but a growing number of heterogeneous Internet-of-Things (IoT) devices, such as smart senors, monitoring devices, home appliances, and autonomous driving vehicles (AVs), for achieving ubiquitous information sharing and seamless communication interaction [1, 2]. The supported service types in 5G also become diverse, ranging from smart homing and remote e-healthcare to intelligent transportation systems and industrial automation [3, 4]. To accommodate increasingly diversified IoT services with growing communication demands, the 5G networks are required to provide customized end-to-end (E2E) service deliveries with different dimensions of quality-of-service (QoS) provisioning. The 3rd generation partnership project (3GPP) technical report (Release 15) defines three main features of 5G networking paradigm with respect to supported service characteristics, connectivity, and QoS requirements [1]:

(1) *Enhanced mobile broadband (eMBB) communications:* The network deployment for 5G needs to be intensified to enhance the overall network capacity for accommodating more and more end user terminals with enhanced mobility (e.g., vehicles on highways, high-speed trains). Moreover, different types of newly emerged data-hungry applications are expected to be supported, such as high-definition (HD) video streaming/conferencing, virtual reality/augmented reality (VR/AR) [5];

© The Author(s), under exclusive license to Springer Nature Switzerland AG 2021
Q. Ye, W. Zhuang, *Intelligent Resource Management for Network Slicing in 5G and Beyond*, Wireless Networks, https://doi.org/10.1007/978-3-030-88666-0_1

(2) *Massive machine-type communication (mMTC) connectivities:* The 5G net-
    works will support a large number of MTC connectivities of heterogeneous
    IoT applications. Different IoT services often have strict and different levels of
    requirements for QoS provisioning. For example, machines in smart factory
    require deterministic timings in processing a sequence of tasks, while the
    latencies for processing/computing autonomous driving tasks are expected to be
    low for guaranteeing prompt AV operations. To this end, the resource utilization
    in 5G needs improvement to boost the network capacity to accommodate the
    ever-increasing IoT traffic;
(3) *Ultra-reliability and low-latency communications (URLLC):* IoT applications
    often have different levels of transmission latency and reliability requirements.
    For example, the information sharing via vehicle-to-vehicle (V2V) commu-
    nications for safety-related applications in intelligent transportation systems
    requires timely information dissemination and high transmission reliability
    [6, 7]. The requirements of processing/transmission latency and reliability for
    autonomous driving tasks become even more strict to ensure prompt and accu-
    rate vehicle operations in response to environment changes [6, 8]. Therefore,
    the IoT service provisioning needs to be customized to guarantee differentiated
    QoS satisfaction and QoS isolation with respect to network dynamics. The
    QoS isolation guarantees minimum levels of service satisfaction which are not
    affected when network dynamics happen due to changes of network traffic load
    conditions, channel quality, and end user mobility [9].

1.1.2   5G Wireless Networks

The key features of 5G services pose significant challenges on providing differentiated QoS for diverse IoT applications through a cost-effective networking
architecture with enhanced resource utilization. In the wireless network domain,
to accommodate massive access from an increasing number of IoT devices with
QoS guarantee, efficient radio resource exploitation is required. The deployment of
multi-layer wireless base stations (BSs) is a potential solution to exploit the spatial
multiplexing gain of currently employed radio spectrum resources. Macro-cell BSs
(MBSs) are deployed to provide wide areas of communication coverages in the first
network tier, which are further underlaid by multi-tiers of small-cell BSs (SBSs)
within the macro-cell coverages. It is also necessary to explore more spectrum
utilization opportunities over unlicensed and underutilized frequency bands through
different wireless access technologies, such as long term evolution (LTE) over
WiFi unlicensed spectrum to support delay-sensitive wireless communication [10],
and mmWave networks operated on ultra-high frequency bands (over 20 GHz)
[11]. However, the spectrum exploitations under current network architecture face
technical challenges: First, the multi-layer BS deployment substantially increases
the capital and operational expenditure (CapEx and OpEx) on establishing dense
network infrastructures, and simultaneously increases the inter-cell interference
levels; Second, the MBSs and SBSs are likely owned by different infrastructure

providers (InPs) [12], with pre-configured radio spectrum resources. Therefore, the spectrum resource sharing among BSs are limited due to large amount of signaling overhead and regulations on InPs.

To facilitate resource sharing among BSs, network function virtualization (NFV) is leveraged to softwarize the higher-layer functions on BSs, where the link-layer and network-layer functionalities are virtualized and decoupled from the physical communication infrastructures and can be flexibly instantiated/placed at different network locations in a cost-effective way [13, 14]. For a 5G wireless network, the NFV infrastructures are established upon a cloud radio access network (C-RAN) architecture, where the baseband processing units (BBUs) are decoupled from each BS and are placed on servers connected the BSs at the edge of the network [15]. In a typical two-tier 5G wireless network architecture, as shown in Fig. 1.1, a main edge server with intensive processing/computing resources in the upper tier is attached to an MBS, and a group of local servers in the lower tier are connected to SBSs,

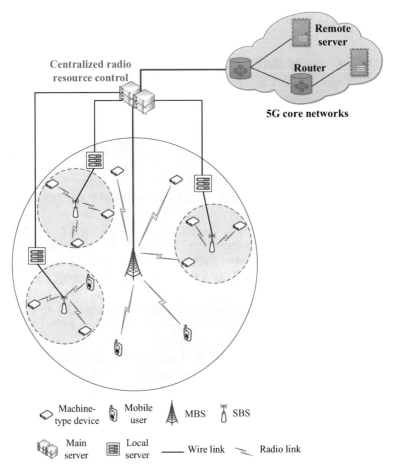

Fig. 1.1 An illustration of two-tier 5G wireless network based on a C-RAN architecture

respectively, to provide lightweight processing capacity near end users/devices. With NFV, the distributed radio resource management (dRRM) functions (e.g., resource scheduling) on BSs are virtualized and migrated to the corresponding servers attached to the BSs and, at the same time, the virtualized central radio resource control (cRRC) function (e.g., radio resource planning) on the MBS is migrated to the main server to coordinate the network-level resource reservation among the BSs. The cRRC is enabled by software-defined networking (SDN) which provides direct programmability on customizing the control policy for the network-wide resource partitioning among the BSs [16, 17]. With SDN-enabled NFV, the radio spectrum resources among BSs can be flexibly reconfigured to enhance the overall resource utilization and dynamically adjusted by adapting to the varying network traffic load. This resource partitioning process is called *radio resource slicing* or *RAN slicing* [4, 18], and the ratio of sliced resources on a BS out of the entire resource pool is termed as *resource slicing ratio* [4]. There is a fundamental research issue on how to optimize the slicing ratios among BSs for maximizing the overall resource utilization with per-user/device QoS satisfaction [9].

1.1.3 5G Core Networks

To realize E2E service deliveries, data traffic generated from the wireless network domain is aggregated into different traffic flows,[1] according to service types at the entry point of a 5G core network. The 5G core network is an interconnection of software-programmable switches/routers (e.g., Open vSwitches [19]) and general-purpose servers via high-speed wired transmission links. Similar to 5G wireless networks, through NFV, network/service functions are decoupled from underlying physical hardware and are flexibly placed on generic servers (also called NFV nodes [20, 21]) at different network locations in order to improve the overall computing resource utilization. Specifically, through resource virtualization platforms (e.g., OpenStack [22]), computing resources (i.e., CPU cores) on servers are virtualized to host different virtual machines (VMs) upon which different network/service functions (e.g., classifiers, firewalls, and transcoding) can be flexibly programmed as software instances referred to as virtual network functions (VNFs). With SDN, the control functionalities on forwarding devices for configuring traffic routing paths are also decoupled and migrated to a centralized network server as a programmable SDN control module. In this way, traffic routing policies are flexibly programmed by the SDN controller for customized E2E service deliveries with differentiated performance satisfaction (e.g., packet delivery ratio, E2E latency, and goodput).

In 5G core networks, a traffic flow belonging to certain application is often required to pass through a specific sequence of network-level or service-level

---

[1] A traffic flow refers to an aggregation of data packets of a same service type being transmitted between two edge points of a 5G core network.

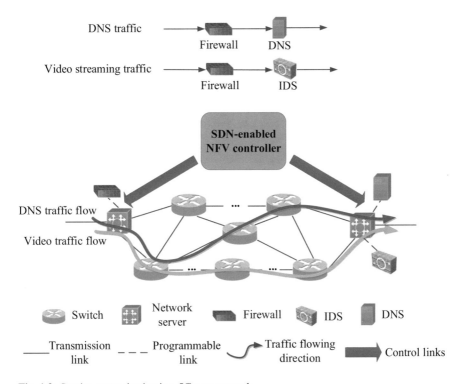

Fig. 1.2 Service customization in a 5G core network

functions to achieve E2E service provisioning, as shown in Fig. 1.2. A set of VNFs with virtual links connecting them constitute a logic VNF chain, also called a *service function chain (SFC)*. Each SFC represents a sequence of VNFs that a traffic flow is required to pass through for E2E service provisioning. For example, each generated video packet from an E2E streaming service needs to traverse a firewall function at the entry point of the core network and then an intrusion detection system (IDS) before being downloaded by a video client to ensure a secure E2E video streaming; To provide a web browsing service, a domain name system (DNS) request is first generated and pass through a firewall function and DNS function to return a secured network domain name to IP address mapping. All VNFs are centrally managed by an NFV controller for efficient function placement, and the controller is also SDN enabled to achieve programmable routing configurations for customized E2E transmission performance. Under the SDN-enabled NFV architecture, increasing numbers/types of VNFs can be flexibly hosted and operated at NFV nodes to realize customized service deliveries and improve the computing and transmission resource utilization without adding more physical resources (e.g., servers, forwarding devices, transmission links).

1.2   Resource Management in 5G Networks

In this section, we discuss how network resources are efficiently managed by introducing the framework of network slicing for both 5G core and wireless networks. From the perspectives of different protocol stack layers, we present the design considerations of virtual network topologies, communication and computing resource slicing, and transport-layer protocol customization.

1.2.1   Network Slicing

Based on the SDN-enabled NFV architecture, in 5G core networks, VNFs are decoupled from hardware-specific servers and virtualized as software instances which can be flexibly instantiated to create different SFCs embedded on a physical substrate for customized service deliveries. To improve the overall resource utilization, it is likely that multiple SFCs share a (partially) common embedded network path, with one or more VNFs placed on a same server and SFC traffic routing paths (partially) overlapped, as shown in Fig. 1.2. When different SFCs share a common physical network path, it is essential to investigate how network resources, including computing resources at NFV nodes and communication resources on transmission links, are properly sliced among traffic flows to enhance the resource utilization and, at the same time, achieve QoS isolation.[2] This process is called *network slicing* [4, 23]. In 5G networks, effective network slicing solutions can be developed by studying the following three research issues from different protocol layer perspectives: (1) In the network planning stage, how to optimize the VNF placement and traffic routing topology for each SFC to minimize the overall CapEx and OpEx, while guaranteeing the QoS isolation among services; (2) How computing and communication resources are jointly sliced among embedded SFCs; (3) In the network operation stage, with the established routing paths and allocated resources, how to customize the transport-layer protocol operated over each embedded SFC to enhance service quality in a more fine-grained manner. In a 5G wireless network which provides the last-hop communications between BSs and end users/devices, the network slicing mainly deals with how to slice radio resources among BSs to enhance the overall communication resource utilization to accommodate increasing groups of service users/devices with QoS isolation, which is called radio resource slicing or RAN slicing [9, 24].

---

[2] QoS isolation refers to any change of network states from one service, due to user mobility, channel quality variations, and traffic load dynamics, should not affect the minimum QoS provided to users/devices from another service type.

1.2.2 Virtual Network Topology

With SDN and NFV, different SFCs can be customized and embedded onto a physical substrate, including VNFs instantiated on appropriate network servers and routing path configuration for traffic passing through the SFCs. This process is referred to as *virtual network topology design*, where an embedded SFC represents a virtual network supporting packet transmission and processing for certain type of service [25, 26]. In the virtual network topology design for embedding SFCs, the VNF placement and traffic routing are correlated, leading to technical challenges on accommodating multiple services over a same physical substrate for efficient resource utilization. Instantiating more NFV nodes for function placement may lead to a reduced link provisioning cost due to balanced data traffic load, but the operational cost for provisioning VNFs on servers can be increased. Therefore, how to balance the tradeoff between VNF and link provisioning is a challenging research issue [27–29]. Our research objective is to determine a joint optimal VNF placement and traffic routing strategy such that the overall function and link provisioning cost can be minimized. Another challenge comes from multiple service customization over a physical network with limited transmission and computing resources. Considering different service characteristics (e.g., data rate, E2E transmission delay), how to embed multiple SFCs has impact on the overall service provisioning cost and service performance. Prioritizing the embedding of a low data-rate SFC fragments the network resources, thus preventing high data-rate services from being successfully embedded [27]. Therefore, in Chap. 2, we develop an efficient framework for the orchestration of multiple services over a common physical substrate, including prioritizing service requests for embedding and jointly optimizing VNF placement and traffic routing for each service to achieve maximum network throughput with minimum SFC provisioning cost. In addition, we establish an accurate analytical model to evaluate the E2E delay of packets traversing each embedded SFC, including packet processing delays on NFV nodes and transmission delays over physical links, with which the delay-aware SFC embedding can be achieved [30].

1.2.3 Resource Slicing

After SFCs are embedded over a physical substrate with some overlapping network paths, the computing resources on NFV nodes and transmission resources on forwarding links need to be properly shared among traffic flows to improve the utilization of both resource types and achieve QoS isolation. Specifically, when a traffic flow of certain service type traverses an NFV node, each packet consumes CPU time for processing and then occupies transmission resources over the outgoing link for packet forwarding. However, different flows may demonstrate discrepant time consumption for CPU processing and link transmission when passing through

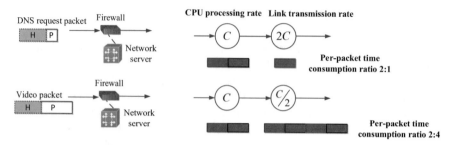

Fig. 1.3 An illustration of bottleneck resource consumption when packets of different traffic flows pass through an NFV node

an NFV node [31, 32]. For example, a DNS request packet for the web browsing service has short data payload size, which consumes more time for CPU processing than link transmission when passing through a firewall function, whereas a video streaming packet occupies more time for link transmission due to a long data payload size. We define *bottleneck resource (or dominant resource)* as the resource type (either computing or transmission resource) that a traffic flow requires more when passing through an NFV node [31]. Figure 1.3 provides an illustration of the time consumption for CPU processing and output link transmission, when packets of different service types traverse a firewall function on a network server. We assume that all computing resources on the server are allocated to process each packet (from either DNS traffic or video streaming traffic) passing through, and C indicates the server CPU processing rate. With the consideration of discrepant bottleneck resource consumption, when multiple traffic flows traverse an NFV node, how to jointly slice both types of resources among flows to achieve high resource utilization with QoS isolation guaranteed and, at the same time, maintain the fairness for resource sharing needs investigation [4].

1.2.4 Customized Protocol

With SDN and NFV, more fine-grained in-network functionalities and control can be enabled for each sliced network to improve the E2E performance (e.g., throughput, delay, and transmission reliability) in the network operation stage. Therefore, it is essential to investigate how the upper protocol stack layer (e.g., transport-layer) functions can be refined with in-network intelligence, for example, in-network congestion detection and early packet loss recovery, to reduce the E2E packet transmission delay. Moreover, the transport-layer protocol functions should be customized for each network slice to provide differentiated service deliveries. For packet-loss tolerant services, e.g., on-demand video streaming, how to balance the trade-off between congestion resolution and end user experience needs investigation [33, 34]; For time-critical services that require high packet transmission reliability and low latency, we investigate how to trigger in-network early packet loss detection and recovery for the purpose of facilitating fast congestion detection and congestion control, such that the E2E delay performance is further enhanced [35, 36]. Therefore, customized slice-level transmission protocols based on the SDN/NFV architecture are expected to be developed with specific protocol functionalities in each network slice for achieving differentiated service provisioning.

1.3 Summary

In this chapter, we have introduced the networking architectures of both 5G wireless and core for supporting diversified services in the IoT era. The concept of network slicing, as an intelligent resource management solution for 5G networks, is presented, which is followed by a discussion on design considerations and potential research issues for developing effective network slicing frameworks from the protocol stack layer perspectives.

References

1. 3rd Generation Partnership Project; Technical specification group services and system aspects; Summary of Rel-15 work items (Release 15), *3GPP TR 21.915 V15.0.0*, Sophia Antipolis Valbonne (2016), pp. 1–31
2. Q. Ye, W. Zhuang, Distributed and adaptive medium access control for Internet-of-Things-enabled mobile networks. IEEE Internet Things J. **4**(2), 446–460 (2017)
3. L. Lyu, C. Chen, S. Zhu, X. Guan, 5G enabled co-design of energy-efficient transmission and estimation for industrial IoT systems. IEEE Trans. Ind. Inf. **14**(6), 2690–2704 (2018)
4. Q. Ye, J. Li, K. Qu, W. Zhuang, X. Shen, X. Li, End-to-end quality of service in 5G networks – Examining the effectiveness of a network slicing framework. IEEE Veh. Technol. Mag. **13**(2), 65–74 (2018)
5. W. Shi, J. Li, P. Yang, Q. Ye, W. Zhuang, X. Shen, Two-level soft RAN slicing for customized services in 5G-and-beyond wireless communications. IEEE Trans. Ind. Inf., to appear. https://doi.org/10.1109/TII.2021.3083579
6. Q. Ye, W. Shi, K. Qu, H. He, W. Zhuang, X. Shen, Joint RAN slicing and computation offloading for autonomous vehicular networks: A learning-assisted hierarchical approach. IEEE Open J. Veh. Technol. **2**, 272–288 (2021)
7. W. Zhuang, Q. Ye, F. Lyu, N. Cheng, J. Ren, SDN/NFV-empowered future IoV with enhanced communication, computing, and caching. Proc. IEEE **108**(2), 274–291 (2020)
8. Q. Ye, W. Shi, K. Qu, H. He, W. Zhuang, X. Shen, Learning-based computing task offloading for autonomous driving: A load balancing perspective, in *Proc. ICC' 21* (2021), pp. 1–6
9. Q. Ye, W. Zhuang, S. Zhang, A. Jin, X. Shen, X. Li, Dynamic radio resource slicing for a two-tier heterogeneous wireless network. IEEE Trans. Veh. Technol. **67**(10), 9896–9910 (2018)
10. N. Zhang, S. Zhang, S. Wu, J. Ren, J. W. Mark, X. Shen, Beyond coexistence: Traffic steering in LTE networks with unlicensed bands. IEEE Wirel. Commun. **23**(6), 40–46 (2016)
11. K. Aldubaikhy, W. Wu, Q. Ye, X. Shen, Low-complexity user selection algorithm for multiuser transmission in mmwaveWLANs. IEEE Trans. Wireless Commun. **19**(4), 2397–2410 (2020)

12. C. Liang, F.R. Yu, H. Yao, Z. Han, Virtual resource allocation in information-centric wireless networks with virtualization. IEEE Trans. Veh. Technol. **65**(12), 9902–9914 (2016)
13. R. Riggio, A. Bradai, D. Harutyunyan, T. Rasheed, T. Ahmed, Scheduling wireless virtual networks functions. IEEE Trans. Netw. Serv. Manag. **13**(2), 240–252 (2016)
14. F. Bari, S.R. Chowdhury, R. Ahmed, R. Boutaba, O.C.M.B. Duarte, Orchestrating virtualized network functions. IEEE Trans. Netw. Serv. Manag. **13**(4), 725–739 (2016)
15. M. Kourtis et al., A cloud-enabled small cell architecture in 5G networks for broadcast/multicast services. IEEE Trans. Broadcast. **65**(2), 414–424 (2019)
16. Q. Duan, N. Ansari, M. Toy, Software-defined network virtualization: An architectural framework for integrating SDN and NFV for service provisioning in future networks. IEEE Netw. **30**(5), 10–16 (2016)
17. A. Belbekkouche, M.M. Hasan, A. Karmouch, Resource discovery and allocation in network virtualization. IEEE Commun. Surv. Tutor. **14**(4), 1114–1128 (2012)
18. W. Wu, N. Chen, C. Zhou, M. Li, X. Shen, W. Zhuang, X. Li, Dynamic RAN slicing for service-oriented vehicular networks via constrained learning. IEEE J. Sel. Areas Commun. **39**(7), 2076–2089 (2021)
19. M. Shahbaz et al., PISCES: A programmable, protocol-independent software switch, in *Proc. ACM SIGCOMM* (2016), pp. 525–538
20. S.Q. Zhang, Q. Zhang, H. Bannazadeh, A. Leon-Garcia, Routing algorithms for network function virtualization enabled multicast topology on SDN. IEEE Trans. Netw. Serv. Manag. **12**(4), 580–594 (2015)
21. L. Wang, Z. Lu, X. Wen, R. Knopp, R. Gupta, Joint optimization of service function chaining and resource allocation in network function virtualization. IEEE Access **4**, 8084–8094 (2016)
22. Openstack (Release Pike). [Online]. Available: https://www.openstack.org. Accessed Dec. 2017
23. X. Shen, J. Gao, W. Wu, K. Lyu, M. Li, W. Zhuang, X. Li, J. Rao, AI-assisted network-slicing based next-generation wireless networks. IEEE Open J. Veh. Technol. **1**, 45–66 (2020)
24. J. Li, W. Shi, P. Yang, Q. Ye, X. Shen, X. Li, J. Rao, A hierarchical soft RAN slicing framework for differentiated service provisioning. IEEE Wireless Commun. **27**(6), 90–97 (2020)
25. J. Li, W. Shi, Q. Ye, S. Zhang, W. Zhuang, and X. Shen, "Joint virtual network topology design and embedding for cybertwin-enabled 6G core networks," *IEEE Internet Things J.*, to be published (DOI: 10.1109/JIOT.2021.3097053).
26. J. Li, W. Shi, Q. Ye, N. Zhang, W. Zhuang, X. Shen, Multiservice function chain embedding with delay guarantee: A game-theoretical approach. IEEE Internet Things J. **8**(14), 11219–11232 (2021)
27. O. Alhussein, P.T. Do, Q. Ye, J. Li, W. Shi, W. Zhuang, X. Shen, X. Li, J. Rao, A virtual network customization framework for multicast services in NFV-enabled core networks. IEEE J. Sel. Areas Commun. **38**(6), 1025–1039 (2020)
28. O. Alhussein, P.T. Do, J. Li, Q. Ye, W. Shi, W. Zhuang, X. Shen, X. Li, J. Rao, Joint VNF placement and multicast traffic routing in 5G core networks, in *Proc. GLOBECOM' 18* (2018), pp. 1–6
29. O. Alhussein, W. Zhuang, Robust online composition, routing and NF placement for NFV-enabled services. IEEE J. Sel. Areas Commun. **38**(6), 1089–1101 (2020)
30. Q. Ye, W. Zhuang, X. Li, J. Rao, End-to-end delay modeling for embedded VNF chains in 5G core networks. IEEE Internet Things J. **6**(1), 692–704 (2019)
31. W. Wang, B. Liang, B. Li, Multi-resource generalized processor sharing for packet processing, in *Proc. ACM IWQoS' 13* (2013, June), pp. 1–10
32. A. Ghodsi, M. Zaharia, B. Hindman, A. Konwinski, S. Shenker, I. Stoica, Dominant resource fairness: Fair allocation of multiple resource types, in *Proc. ACM NSDI' 11* (2011, April), pp. 24–37
33. S. Yan, P. Yang, Q. Ye, W. Zhuang, X. Shen, X. Li, J. Rao, Transmission protocol customization for network slicing: A case study of video streaming. IEEE Veh. Technol. Mag. **14**(4), 20–28 (2019)

34. S. Yan, Q. Ye, W. Zhuang, Learning-based transmission protocol customization for VoD streaming in Cybertwin-enabled next generation core networks. IEEE Internet Things J., to be published. https://doi.org/10.1109/JIOT.2021.3097628
35. J. Chen, Q. Ye, W. Quan, S. Yan, P.T. Do, P. Yang, W. Zhuang, X. Shen, X. Li, J. Rao, SDATP: An SDN-based traffic-adaptive and service-oriented transmission protocol. IEEE Trans. Cogn. Commun. **6**(2), 756–770 (2020)
36. J. Chen, Q. Ye, W. Quan, S. Yan, P.T. Do, W. Zhuang, X. Shen, X. Li, J. Rao, SDATP: An SDN-based adaptive transmission protocol for time-critical services. IEEE Netw. **34**(3), 154–162 (2020)

Chapter 2
Virtual Network Topology Design and E2E Delay Modeling

2.1 Multi-Service Virtual Network Topology Design

For future applications, there is a growing demand for services of multicast nature, for instance, HD video streaming/conferencing, AR of multiple players, and file downloading. To accommodate multicast services in a 5G core network, multiple instances of one VNF may need to be instantiated at different network server locations, such that the traffic flows are routed through different paths, traversing geographically dispersed VNF instances, to reach different multicast destinations. Figure 2.1 provides an example of multicast service function chaining for a video conferencing service that distributes video contents to different destinations. With the emergence of SDN, a global view of the physical substrate is provided, which makes Steiner tree-based routing approaches feasible [1–3]. However, these routing approaches do not consider the VNF placement in the problem formulation.

To customize a virtual network for E2E service delivery, the VNF placement and multicast traffic routing problems need to be jointly studied. However, due to the correlation between VNF placement and traffic routing, it is technically challenging to even embed a single multicast service over a physical substrate: (1) Choosing a small number of NFV nodes for VNF placement increases the link provisioning cost for establishing a multicast routing topology; (2) On the contrary, instantiating VNF instances on more NFV nodes achieves a balanced traffic routing strategy with increased VNF provisioning cost. Therefore, we intend to investigate how to strike a balance on link and function provisioning such that the total cost for accommodating a multicast service over the physical network is minimized. Another challenge comes from multiple multicast service provisioning on a shared physical substrate. With limited transmission and computing resources on the physical network, prioritizing the SFC embedding for a low-data-rate service can fragment the network resources, which prevents other high-data-rate services from being effectively accommodated and leads to a low overall resource utilization. Some existing studies address the multicast provisioning problem by minimizing the

Q. Ye, W. Zhuang, *Intelligent Resource Management for Network Slicing in 5G and Beyond*, Wireless Networks, https://doi.org/10.1007/978-3-030-88666-0_2

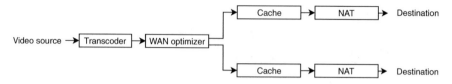

Fig. 2.1 An illustration of a multicast SFC for a video conferencing service

function and link provisioning costs [4–9]. However, most proposals assume that all VNFs from each service are placed on one NFV node and multicast replication points only occur after the placement of VNFs. A more realistic design of VNF placement and traffic routing needs investigation, e.g., flexibly instantiating VNFs on different NFV nodes with multi-path traffic routing between functions.

In this section, we establish an optimization framework for the multiple multicast service provisioning over a physical substrate. We first study a joint VNF placement and traffic routing problem for a single service provisioning problem to minimize the function and link provisioning costs. For practicality, we consider *one-to-many* and *many-to-one* mappings between VNFs and NFV nodes to achieve a flexible multicast traffic routing and VNF placement, where multiple VNF instances can be hosted on one NFV node and one type of VNF can also be replicated and placed on different NFV nodes. Moreover, we incorporate in the problem formulation the cases of single-path and multi-path traffic routing between embedded VNF instances. As the formulated problem is NP-hard, a low-complexity heuristic algorithm is devised to determine an efficient solution. For a general scenario of placing multiple services over the common physical network, the service requests compete with each other to be hosted over the substrate network with resource constraints. Therefore, we consider to accept the services achieving maximum throughput with minimum function and link provisioning costs. The problem is formulated as an mixed integer linear program (MILP) that jointly determines the virtual network topologies for the embedded services. A simple heuristic algorithm is designed for prioritizing the service requests for the SFC embedding.

2.1.1 Physical Substrate Network

Consider a physical substrate network, denoted by $\mathcal{G} = (\mathcal{N}, \mathcal{L})$, where $\mathcal{N}$ and $\mathcal{L}$ are the sets of network nodes and transmission links, respectively. The network nodes can be forwarding devices, represented by $\mathcal{F}$, and NFV nodes, represented by set $\mathcal{M}$, i.e., $\mathcal{N} = \mathcal{F} \cup \mathcal{M}$. The forwarding devices are switches/routers capable of forwarding and replicating traffic, and NFV nodes are network servers hosting VNFs. Assume that each NFV is configured with the CPU processing rate of $C(n), n \in \mathcal{M}$, in packet per second (packet/s) [10–12], and also has traffic forwarding capability. Moreover, multiple VNFs can be provisioned on an NFV node simultaneously as long as the available processing rate meets the function processing requirements. Each physical link has a transmission rate of $B(l)$, $l \in \mathcal{L}$ (in packet/s).

2.1.2 Multicast SFCs

The types of VNFs supported in the network are represented by set $\mathcal{P}$, where a specific type indicates certain virtualized functionality (e.g., firewall, transcoding, DNS, and proxy). We further use $k_{ni} \in \{0, 1\}$ to indicate whether virtual function $f_i (\in \mathcal{P})$ can be instantiated on NFV node $n (\in \mathcal{M})$, where $k_{ni} = 1$ if f_i is admittable to NFV node n and $k_{ni} = 0$ otherwise.

The set of multicast service requests are denoted by $\mathcal{R}$. We describe the multicast SFC of service $r(\in \mathcal{R})$ as a weighted acyclic directed graph, given by

$$S^r = (s^r, \mathcal{D}^r, \mathcal{V}^r, \overline{d^r}), \quad r \in \mathcal{R} \tag{2.1}$$

where s^r and $\mathcal{D}^r$ denote the source and the set of multicast destination nodes, respectively, $\mathcal{V}^r = \{f_1^r, f_2^r, \ldots f_{|\mathcal{V}|}^r\}$ represents a set of VNFs that need to be passed through sequentially for each source-destination node pair, and $\overline{d^r}$ is the data rate requirement (in unit of packet/s) of service r. Each VNF f_i^r has its packet processing rate requirement $C(f_i^r)$ (in packet/s), $i \in \{1, \ldots, |\mathcal{V}^r|\}$. Each VNF instance belongs to one service, which cannot be shared among multiple SFCs [5, 6, 9, 13].

2.1.3 Problem Formulation

We first investigate a joint multicast traffic routing and VNF placement problem for embedding a single service over a physical substrate. Given the substrate network and the multicast service description, we aim at designing an embedded virtual network topology for delivering the multicast service on the physical network, including (1) VNF placement on different NFV nodes, and (2) multicast traffic routing path configuration from the source node to the multicast destination nodes by passing through a sequence of VNFs. The multipath routing between VNFs is enabled to improve the network resource utilization and provide the flexibility for routing topology design. Our objective is to minimize the SFC embedding cost for VNF and link provisioning for determining an optimal embedded virtual network topology. However, as discussed in Sect. 2.1, instantiating a large number of VNF instances at different server locations balances the network traffic load with increased function provisioning cost, whereas fewer VNF instantiations saves VNF placement cost at the expense of imbalanced load and less efficient resource utilization. Therefore, it is required to balance the tradeoff between VNF instantiation and transmission link provisioning.

For the second problem, we intend to study a joint multicast routing and VNF placement problem in a multi-service scenario. Due to the limited network resources, it is required to determine which SFCs need to be embedded over the physical network such that the aggregate throughput is maximized with minimum function and link provisioning costs. A static SFC embedding is considered, where the SFCs from all service requests are ready to be embedded a priori. The objective is to find an optimal combination of multiple virtual network topologies for multicast SFC embedding to achieve maximum throughput, while minimizing the SFC provisioning costs.

We consider the one-to-many and many-to-one mapping between VNF intances/virtual links and NFV nodes/physical links, which facilitates a flexible and efficient SFC embedding for largely distributed networks where destination nodes are geographically dispersed. An efficient way is to deploy duplicated VNF instances close to destinations for flexible routing. Figure 2.2 illustrates how a multicast SFC with two VNFs and two destination nodes (i.e., $t_1, t_2 \in \mathcal{D}$) is embedded onto the physical network with resultant virtual topologies, considering both non-flexible and flexible function placement. It can be seen that with multiple instances of VNF f_2 placed close to destinations, as shown in Fig. 2.2d, the flexibly

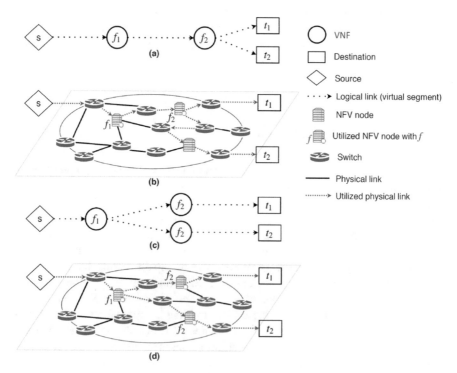

Fig. 2.2 A comparison of flexible and non-flexible multicast SFC embedding. (**a**) SFC topology; (**b**) Non-flexible SFC Embedding; (**c**) Modified SFC topology; (**d**) Flexible SFC embedding

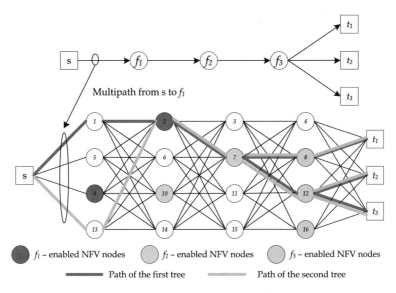

Fig. 2.3 Two Steiner trees that share the same source, traversed functions, and destinations

designed virtual network topology uses less number of transmission links for traffic routing compared with the non-flexible case in Fig. 2.2b.

2.1.3.1 Multipath-Enabled Single Service Scenario

Denote by J^r the maximum number of established tree topologies to route traffic from source to destinations for multicast service r ($\in \mathcal{R}$). As illustrated in Fig. 2.3, each multicast routing tree emanates from the source and traverses a same set of VNFs placed on NFV nodes to reach the destinations. In Fig. 2.3, we have one multicast service request with three VNFs and three destination nodes, and we use two multipath routing trees ($J^r = 2$) to support the service delivery. It can be seen that the multipath routes from the source node to the function f_1 at node 2 and converges afterwards. The maximum number of established multicast traffic routing trees supported by the physical substrate, J^r, is an input parameter in our problem formulation. In the following, we present our joint VNF placement and multipath-enabled traffic routing formulation for a single service case.

Let $\mathcal{S}_m^n$ denote the set of integers from m to n ($> m$), i.e., $\mathcal{S}_m^n \triangleq \{m, m+1, \ldots, n\}$ with $m, n \in \mathbf{Z}_+$. We define binary variable $x_{li}^{jr} \in \{0, 1\}$, where $x_{li}^{jr} = 1$ indicates that link l ($\in \mathcal{L}$) is selected for directing traffic from function f_i^r to function f_{i+1}^r ($i \in \mathcal{S}_1^{|\mathcal{V}^r|-1}$) for service r in multicast tree j, binary variable $x_{l0}^{jr} = 1$ indicates that link l forwards traffic from s^r to function f_1^r, and $x_{l|\mathcal{V}|}^{jr} = 1$ indicates that link l directs traffic from function $f_{|\mathcal{V}|}^r$ to any destination. We define $y_{lit}^{jr} \in \{0, 1\}$, where

$y_{lit}^{jr} = 1$ denotes that link l forwards traffic for service r in multicast tree j from VNF f_i^r to VNF f_{i+1}^r for destination t ($\in \mathcal{D}^r$), binary variable $y_{l0t}^{jr} = 1$ indicates that link l directs traffic for service r in tree j from s^r to VNF f_1^r for destination t, and $y_{l|\mathcal{V}|t}^{jr} = 1$ indicates that link l directs traffic for service r in tree j from VNF $f_{|\mathcal{V}|}^r$ to destination t. By defining $\boldsymbol{x} = \{x_{li}^{jr}\}$ and $\boldsymbol{y} = \{y_{lit}^{jr}\}$, we have

$$y_{lit}^{jr} \leq x_{li}^{jr}, \quad l \in \mathcal{L}, i \in \mathcal{S}_0^{|\mathcal{V}^r|}, j \in \mathcal{S}_1^{J^r}, t \in \mathcal{D}^r, r \in \mathcal{R}. \tag{2.2}$$

Also, we define binary variable $z_{ni}^r \in \{0, 1\}$ as an indicator for the deployment of function f_i^r ($i \in \mathcal{S}_1^{|\mathcal{V}|}$) on NFV node n for service r, where $z_{ni}^r = 1$ indicates the placement of f_i^r on NFV node n and $z_{ni}^r = 0$ otherwise. The binary variable u_{nit}^r is also defined to indicate the placement status of f_i^r on NFV node n for service r for destination t (i.e., $u_{nit}^r = 1$ indicates the placement and $u_{nit}^r = 0$ otherwise). Similarly, the relation constraint between $\boldsymbol{z} = \{z_{ni}^r\}$ and $\boldsymbol{u} = \{u_{nit}^r\}$ is given by

$$u_{nit}^r \leq z_{ni}^r, \quad n \in \mathcal{N}, i \in \mathcal{S}_1^{|\mathcal{V}^r|}, t \in \mathcal{D}^r, r \in \mathcal{R}. \tag{2.3}$$

For multicast service r ($\in \mathcal{R}$), we establish J^r multicast trees for multipath traffic routing, and each tree contributes a fraction of transmission rate d^{jr} in order to satisfy the overall required data rate $\overline{d^r}$. Thus, the follow constraint is imposed, which is given by

$$\sum_{j=1}^{J^r} d^{jr} \geq \overline{d^r}, r \in \mathcal{R}. \tag{2.4}$$

Next, the traffic routing and VNF placement constraint is also incorporated as given in (2.5) to ensure traffic flows traversing the SFC from source to multiple destinations. Denote f_0^r and $f_{|\mathcal{V}|+1}^r$ as dummy VNFs (with no computing requirements) placed on the source node s^r and each destination node, respectively. As some of the multicast routing trees can be deactivated, we have

$$\sum_{(n,m)\in\mathcal{L}} y_{(n,m)it}^{jr} - \sum_{(m,n)\in\mathcal{L}} y_{(m,n)it}^{jr} = \begin{cases} u_{nit}^r - u_{n(i+1)t}^r, & \text{activate tree } j \\ 0, & \text{otherwise} \end{cases} \tag{2.5}$$

for $n \in \mathcal{N}, i \in \mathcal{S}_0^{|\mathcal{V}^r|}, t \in \mathcal{D}^r, r \in \mathcal{R}$, where

$$u_{s0t}^r = 1, u_{n0t}^r = 0, \forall t \in \mathcal{D}^r, n \in \mathcal{N}\backslash\{s^r\}$$

$$u_{t(|\mathcal{V}^r|+1)t} = 1, u_{n(|\mathcal{V}^r|+1)t} = 0, \forall t \in \mathcal{D}^r, n \in \mathcal{N}\backslash\mathcal{D}^r$$

$$u_{sit}^r = 0, u_{tit}^r = 0, \forall i \in \mathcal{S}_1^{|\mathcal{V}^r|}, t \in \mathcal{D}^r.$$

Define binary variable $\pi^{jr} \in \{0, 1\}$ which indicates the activation of tree j of service r when $\pi^{jr} = 1$. Otherwise, variables x_{li}^{jr}, y_{lit}^{jr}, and d^{jr} related to deactivated trees are forced to 0. Therefore, we impose the constraints given in (2.6).

$$x_{li}^{jr} \leq \pi^{jr}, \ y_{lit}^{jr} \leq \pi^{jr}, d^{jr} \leq \pi^{jr}\overline{d^r}, \ l \in \mathcal{L}, i \in \mathcal{S}_0^{|\mathcal{V}^r|}, j \in \mathcal{S}_1^{J^r}, t \in \mathcal{D}^r, r \in \mathcal{R}. \tag{2.6}$$

To satisfy the VNF placement and traffic routing constraints only when tree j from service r is activated, Eq. (2.5) is modified as

$$\sum_{(n,m) \in \mathcal{L}} y_{(n,m)it}^{jr} - \sum_{(m,n) \in \mathcal{L}} y_{(m,n)it}^{jr} = \pi^{jr}\left(u_{n(i+1)t}^r - u_{nit}^r\right),$$

$$n \in \mathcal{N}, i \in \mathcal{S}_0^{|\mathcal{V}^r|}, t \in \mathcal{D}^r, r \in \mathcal{R}. \tag{2.7}$$

As $y_{lit}^{jr} \leq x_{li}^{jr}$ is incorporated in (2.2), Eq. (2.6) is rewritten as (2.8) with the constraint $y_{lit}^{jr} \leq \pi^{jr}$ omitted.

$$x_{li}^{jr} \leq \pi^{jr}, \ d^{jr} \leq \pi^{jr}\overline{d^r}, \ l \in \mathcal{L}, i \in \mathcal{S}_0^{|\mathcal{V}^r|}, j \in \mathcal{S}_1^{J^r} \tag{2.8}$$

where x_{li}^{jr} and d^{jr} are considered when tree j is activated; otherwise, they are set to zero. It is also required that exactly one instance of VNF f_i^r is passed through for every source-destination node pair, which is expressed as

$$\sum_{n \in M} u_{nit}^r = 1, \ i \in \mathcal{S}_1^{|\mathcal{V}^r|}, t \in \mathcal{D}^r, r \in \mathcal{R}. \tag{2.9}$$

Moreover, VNF f_i^r can be instantiated at node n only when the function is allowed to be hosted and the computing resources at node n are sufficient. Thus, we have

$$\sum_{r \in \mathcal{R}} \sum_{i=1}^{|\mathcal{V}^r|} z_{ni}^r C(f_i^r) \leq C(n), \ n \in M, \ i \in \mathcal{S}_1^{|\mathcal{V}^r|} \tag{2.10a}$$

$$z_{ni}^r k_{ni} = 1, \ n \in M, i \in \mathcal{S}_1^{|\mathcal{V}^r|}, r \in \mathcal{R} \tag{2.10b}$$

where k_{ni} is an indicator of whether NFV node n admits function f_i. $k_{ni} = 1$ if f_i is admitted on node n; otherwise, $k_{ni} = 0$.

Research Objective We aim at minimizing the overall function and link provisioning costs over J^r multicast trees for service r to balance the network resource utilization, which is expressed as

$$\min \alpha \sum_{l \in \mathcal{L}} \sum_{j=1}^{J^r} \sum_{i=0}^{|\mathcal{V}^r|} \left(\frac{d^{jr}}{B(l)} + 1\right) x_{li}^{jr} + \beta \sum_{i=1}^{|\mathcal{V}^r|} \sum_{n \in M} \frac{C(f_i^r)}{C(n)} z_{ni}^r. \tag{2.11}$$

In (2.11), the objective function is a weighted summation of the cost of traffic forwarding over the selected physical transmission links for all activated trees, and the cost of VNF instance provisioning in the NFV nodes. Parameters α and β are the weighting factors to balance the importance between minimizing the traffic forwarding cost and minimizing the VNF provisioning cost, where $\alpha + \beta = 1$ and $\alpha, \beta > 0$. The terms $d^{jr} x_{li}^{jr}/B(l)$ and $C(f_i^r)z_{ni}^r/C(n)$ is to achieve load balancing over transmission links and NFV nodes [14]. Moreover, the term '+1' in (2.11) is used to minimize the number of links in establishing the multicast topology. Denote by γ_{li}^{jr} the product of d^{jr} and x_{li}^{jr} in the objective function (2.11), given by

$$\gamma_{li}^{jr} = x_{li}^{jr} d^{jr}, \ l \in \mathcal{L}, i \in S_0^{|\mathcal{V}^r|}, j \in S_1^{J^r}, r \in \mathcal{R} \tag{2.12}$$

where γ_{li}^{jr} is interpreted as the transmission rate over link l to forward traffic from function f_i^r to function f_{i+1}^r in tree j. The aggregate rate over link l from all services is upper bounded by the available link transmission rate $B(l)$, i.e.,

$$\sum_{r \in \mathcal{R}} \sum_{j=1}^{J^r} \sum_{i=0}^{|\mathcal{V}^r|} \gamma_{li}^{jr} \leq B(l), \ l \in \mathcal{L}. \tag{2.13}$$

Therefore, the joint VFN placement and multicast traffic routing problem for a single service scenario is formulated as

$$\textbf{(P1)}: \ \min \alpha \sum_{l \in \mathcal{L}} \sum_{j=1}^{J^r} \sum_{i=0}^{|\mathcal{V}^r|} \left(\frac{\gamma_{li}^{jr}}{B(l)} + x_{li}^{jr} \right) + \beta \sum_{i=1}^{|\mathcal{V}^r|} \sum_{n \in M} \frac{C(f_i^r)}{C(n)} z_{ni}^r \tag{2.14a}$$

$$\text{subject to} \quad (2.2) - (2.4), (2.7) - (2.10), (2.12), (2.13) \tag{2.14b}$$

$$\boldsymbol{x}, \boldsymbol{y}, \boldsymbol{z}, \boldsymbol{u}, \boldsymbol{\pi} \in \{0, 1\}, \ \boldsymbol{d} \geq 0, \ \boldsymbol{\gamma} \geq 0. \tag{2.14c}$$

where we have linear objective function and constraints except constraints (2.7), (2.12), and (2.10b). To solve (P1), the non-linear constraints are transformed to equivalent linear constraints such that the problem is converted to an MILP. Specifically, the bilinear term $\pi^{jr} u_{nit}^r$ in constraint (2.7) is handled by introducing a new variable $w_{nit}^{jr} = \pi^{jr} u_{nit}^r$ and the constraint (2.7) is changed to

$$\sum_{(m,n) \in \mathcal{L}} y_{(m,n)it}^{jr} - \sum_{(n,m) \in \mathcal{L}} y_{(n,m)it}^{jr} = w_{nit}^{jr} - w_{n(i+1)t}^{jr},$$

$$n \in \mathcal{N}, i \in S_0^{|\mathcal{V}^r|}, t \in \mathcal{D}^r, j \in S_1^{J^r}, r \in \mathcal{R}. \tag{2.15}$$

The corresponding relations among w_{nit}^{jr}, π^{jr}, and u_{nit}^{r} are given by

$$w_{nit}^{jr} \leq \pi^{j}, \; w_{nit}^{jr} \leq u_{nit}^{r}, \; w_{nit}^{jr} \geq \pi^{jr} + u_{nit}^{r} - 1,$$

$$n \in \mathcal{N}, i \in \mathcal{S}_{0}^{|\mathcal{V}^{r}|}, t \in \mathcal{D}^{r}, j \in \mathcal{S}_{1}^{J^{r}}, r \in \mathcal{R}. \tag{2.16}$$

We denote the product term $z_{ni}^{r} k_{ni}$ by g_{ni}^{r} in nonlinear constraint (2.10b), and consequently, (2.10b) is expressed as

$$g_{ni}^{r} \leq z_{ni}^{r}, \; n \in \mathcal{M}, \; i \in \mathcal{S}_{1}^{|\mathcal{V}^{r}|} \tag{2.17a}$$

$$g_{ni}^{r} \leq k_{ni}, \; n \in \mathcal{M}, \; i \in \mathcal{S}_{1}^{|\mathcal{V}^{r}|} \tag{2.17b}$$

$$g_{ni}^{r} \geq z_{ni}^{r} + k_{ni} - 1, \; n \in \mathcal{M}, \; i \in \mathcal{S}_{1}^{|\mathcal{V}^{r}|}. \tag{2.17c}$$

The big-M method is utilized for nonlinear constraint (2.12), which is expressed equivalently as

$$d^{jr} - M(1 - x_{li}^{jr}) \leq \gamma_{li}^{jr} \leq d^{jr}, \; l \in \mathcal{L}, i \in \mathcal{S}_{0}^{|\mathcal{V}^{r}|}, \; j \in \mathcal{S}_{1}^{J^{r}}, \; r \in \mathcal{R} \tag{2.18a}$$

$$0 \leq \gamma_{li}^{jr} \leq M x_{li}^{jr}, \; l \in \mathcal{L}, i \in \mathcal{S}_{0}^{|\mathcal{V}^{r}|}, \; j \in \mathcal{S}_{1}^{J^{r}}, \; r \in \mathcal{R} \tag{2.18b}$$

where M is a large positive number. As $\overline{d^{r}}$ is the upper limit of d^{jr}, γ_{li}^{jr} in (2.12) is upper bounded by $\overline{d^{r}}$. Thus, it is feasible to set $M = \overline{d^{r}}$.

Consequently, problem ((P1)) for a single-service case is transformed into an MILP form as

$$(\mathbf{P1'}): \; \min_{\mathcal{X}} \sum_{l \in \mathcal{L}} \sum_{j=1}^{J^{r}} \sum_{i=0}^{|\mathcal{V}^{r}|} \alpha \left(\frac{\gamma_{li}^{jr}}{B(l)} + x_{li}^{jr} \right) + \beta \sum_{i=1}^{|\mathcal{V}^{r}|} \sum_{n \in \mathcal{M}} \frac{C(f_{i}^{r})}{C(n)} z_{ni}^{r} \tag{2.19a}$$

$$\text{subject to } (2.2) - (2.4), (2.8) - (2.10a), (2.13), (2.14c) - (2.18) \tag{2.19b}$$

where $\mathcal{X} = \{x, y, z, u, w, \pi, d, \gamma\}$, and an MILP solver is used to deal with problem (P1').

2.1.3.2 Multipath-Enabled Multi-Service Scenario

In this subsection, the problem of embedding multiple multicast services is studied. An MILP is established to construct the multicast topology for multiple SFCs. The overall objective is to find a combination of SFCs to be embedded over a physical network such that the network throughput is maximized while the VNF and link provisioning costs are minimized. The achieved aggregate network throughput is given by

$$\mathbb{R} = \sum_{r \in \mathcal{R}} R^r \rho^r \tag{2.20}$$

where $\rho^r (\in \{0, 1\})$ is a binary decision variable with $\rho^r = 1$ when service r is accepted, and R^r is the achieved throughput for service r, expressed as

$$R^r = a_1 \sum_{i=1}^{|\mathcal{V}^r|} C(f_i^r) + a_2 \left(|\mathcal{V}^r| + |\mathcal{D}^r|\right) \overline{d^r}, \ r \in \mathcal{R} \tag{2.21}$$

where the first and second terms indicate the required computing and transmission rates (in packet/s), respectively [10–12]. The parameters, a_1 and a_2, are the priority weighting factors of computing and transmission rates respectively, where $a_1 + a_2 = 1$ and $a_1, a_2 > 0$.

In addition to maximizing the network throughput, an efficient SFC configuration is needed to minimize the function and link provisioning cost to save resources for future services. The multi-service scenario is considered by formulating a two-step MILP. The first step aims at achieving maximum aggregate network throughput, followed by determining joint traffic routing and VNF placement for each admitted SFC subject to the maximum achievable aggregate throughput.

In a multi-service scenario, some service requests may not be admitted due to resource limitations. Therefore, some of the previous constraints are generalized in the following. We generalize constraint (2.4) to

$$\sum_{j=1}^{J^r} d^{jr} \geq \rho^r \overline{d^r}, \ r \in \mathcal{R} \tag{2.22}$$

where the summation of the fractional transmission rates from all trees for service r ($\in \mathcal{R}$) becomes zero when the service is rejected (i.e., $\rho^r = 0$). Moreover, a VNF instance of f_i^r is deployed at only one NFV node if service r is accepted, i.e., (2.9) becomes

$$\sum_{n \in M} u_{nit}^r = \rho^r, \ i \in \mathcal{S}_1^{|\mathcal{V}^r|}, \ t \in \mathcal{D}^r, \ r \in \mathcal{R}. \tag{2.23}$$

Similarly, when service r is rejected, all variables for the service become zero, i.e.,

$$\pi^{jr} \leq \rho^r, \ z_{ni}^r \leq \rho^r, \ n \in \mathcal{N}, i \in \mathcal{S}_1^{|\mathcal{V}^r|}, \ j \in \mathcal{S}_1^{J^r}, \ r \in \mathcal{R}. \tag{2.24}$$

For the first step, the objective is to determine the set of admitted services that maximize the network throughput $\mathbb{R}^*$; For the second step, we aim at minimizing the function and link provisioning costs for all admitted services, subject to the maximum achievable network throughput $\mathbb{R}^*$. Specifically, the problem of maximizing the network throughput is presented as

$$\textbf{(P2) :} \quad \max_{x,y,z,u,\rho,\pi,w,d,r} \sum_{r \in \mathcal{R}} R^r \rho^r \tag{2.25a}$$

subject to

$$(2.2), (2.3), (2.8), (2.10a), (2.13), (2.15), (2.16) \tag{2.25b}$$

$$(2.18), (2.17), (2.22) - (2.24) \tag{2.25c}$$

$$x, y, z, u, \rho, \pi, w \in \{0, 1\}, \ d, r \geq 0. \tag{2.25d}$$

After solving (2.25), we obtain a configuration that achieves maximum network throughput, $\mathbb{R}^*$, over the substrate network, given $|\mathcal{R}|$ services. Among all possible configurations, we also need to select one with which the function and link provisioning costs are minimized.

Therefore, in the second step, we determine the combination of embedded services and their multicast virtual network topologies that leads to the minimum function and link provisioning costs, subject to the maximum achievable network throughput $\mathbb{R}^*$, by solving (P3).

$$\textbf{(P3) :} \quad \min_{x,y,z,u,\rho,\pi,w,d,r} \sum_{r \in \mathcal{R}} \sum_{l \in \mathcal{L}} \sum_{j \in \mathcal{S}_1^{Jr}} \sum_{i=0}^{|\mathcal{V}^r|} \alpha \left(\frac{\gamma_{li}^{jr}}{B(l)} + x_{li}^{jr} \right) + \sum_{r \in \mathcal{R}} \sum_{i=1}^{|\mathcal{V}^r|} \sum_{n \in \mathcal{N}} \beta \frac{C(f_i^r)}{C(n)} z_{ni}^r$$

subject to

$$(2.2), (2.3), (2.8), (2.10a), (2.17), (2.13), (2.15), (2.16), (2.18), (2.22) - (2.24)$$

$$x, y, z, u, \rho, \pi, w \in \{0, 1\}, \ d, r \geq 0 \tag{2.26a}$$

$$\sum_{r \in \mathcal{R}} R^r \rho^r \geq \mathbb{R}^*. \tag{2.26b}$$

The problem in (2.26) is an MILP, which can be solved using a standard MILP solver, and the optimal solutions achieve maximum network throughput $\mathbb{R}^*$ with minimum function and link provisioning costs.

Remark 2.1 The joint multicast traffic routing and VNF placement are NP-hard problems for both single-service and multi-service cases.

Due to space limit, the proof for *Remark 2.1* is omitted which can be found in [15].

2.1.4 Heuristic Solution

As the time complexity in solving (P1$'$), (P2), and (P3) in Sect. 2.1.3 is high when using an MILP solver, a low-complexity heuristic algorithm is needed to find solutions more efficiently. The solution framework is proposed in two steps: (1) service requests are prioritized based on heuristics that maximize the aggregate network throughput, and (2) the prioritized service requests are embedded sequentially using the joint placement and routing (JPR) algorithm for each single service.

2.1.4.1 Single-Service Scenario

A single-service heuristic algorithm is designed based on the following considerations: (1) Different types of VNFs are able to run simultaneously on an NFV node; (2) The VNF traversing order in an SFC should be satisfied for each source-destination (S-D) pair; (3) The objective is to minimize the provisioning cost of the multicast virtual network topology based on (2.11). A two-step heuristic algorithm is devised as follows: First, the link provisioning cost is minimized by constructing a key-node preferred MST (KPMST)-based multicast topology that connects the source with the destinations; Then, we greedily perform VNF placement to minimize the number of VNF instances. The pseudo-code of the JPR heuristic is shown in Algorithm 1, which is explained in more detail as follows.

First, the substrate network G is copied into G'. Second, we calculate new link weights for G' to prioritize the NFV node selection in building the KPMST, given by

$$\omega_{l'} = \alpha \left(\frac{\overline{d^r}}{B(l')} + 1 \right) + \beta \frac{\overline{d^r}}{C(m)}, \quad l' = (n, m) \in \mathcal{L}', \mathcal{L}' \subseteq G', r \in \mathcal{R} \qquad (2.27)$$

where $C(m)$ is set to a small value when m is a network switch; otherwise, it represents the processing resource of NFV node m. A key-NFV node is chosen iteratively. We construct the metric closure of G' as G'', which is a complete weighted graph with the same node set N and with the new link weight over any pair of nodes given by the respective shortest path distance. From the metric closure, we find the MST which connects the source node, destination nodes, and the key-NFV node. A multicast traffic routing topology (G_v) is initialized by replacing the edges in the MST with corresponding paths in G' wherever needed. Then, the VNFs are greedily placed from the source of the multicast topology to its destinations with the objective of minimizing the number of VNF instances. The cost $\mathbb{C}(G_v)$ of the new multicast topology (as in (2.11)) and the number $\mathbb{A}(G_v)$ of successfully embedded VNF instances are computed. Our objective is to jointly maximize the number of successfully admitted VNFs and minimize the provisioning cost by iterating over all candidate key-NFV nodes. In every iteration, a new key-NFV node is chosen.

Algorithm 1: Heuristic algorithm for the joint VNF placement and routing

---

1 Procedure JPR ($\mathcal{G}, S^r$);

   Input : $\mathcal{G}'(\mathcal{N}', \mathcal{L}'), S^r = (s^r, \mathcal{D}^r, f_1^r, f_2^r, \ldots, f_{|\mathcal{V}|}^r, \overline{d^r})$

   Output: $\mathcal{G}_v$

2 $\mathbb{C}_r \leftarrow \infty$; $\mathbb{A}_r \leftarrow 0$;

3 **for** $n \in \mathcal{M}$ **do**

4    $\mathcal{G}'' \leftarrow \text{MetricClosure}(\mathcal{G}', \{n, s, \mathcal{D}\}); \mathcal{G}_v^{temp}(\mathcal{N}_v, \mathcal{L}_v) \leftarrow \text{KruskalMST}(\mathcal{G}'')$;

5   **for** *path from s to each* $t \in \mathcal{D}$ **do** place VNFs from $\mathcal{V}$ on available NFV nodes sequentially subject to (2.10);

6   **if** $\mathbb{A}(\mathcal{G}_v^t) = \mathbb{A}_r$ & $\mathbb{C}(\mathcal{G}_v^t) < \mathbb{C}_r$ **then** $\mathcal{G}_v \leftarrow \mathcal{G}_v^t$; $\mathbb{C}_r \leftarrow \mathbb{C}(\mathcal{G}_v^t)$;

7   **else if** $\mathbb{A}(\mathcal{G}_v^t) > \mathbb{A}_{ref}$ **then** $\mathcal{G}_v \leftarrow \mathcal{G}_v^t$; $\mathbb{A}_{ref} \leftarrow \mathbb{A}(\mathcal{G}_v^t)$; $\mathbb{C}_{ref} \leftarrow \mathbb{C}(\mathcal{G}_v^t)$;

8 **end**

9 **for** *each missing VNF* (f) *from* $\mathcal{G}_v$ **do** link nearest NFV node that can host f to $P_{c,t,f}$, and remove unnecessary edges;

10 **for** *path from* f_i *to* f_{i+1} *for each* t *in* $\mathcal{G}_v$ **do**

11   success $\leftarrow$ false;

12   **for** $j = 1 : J$ **do**

13   Find $temp = \min\{K_{i,i+1}^t, j\}$ paths from $\mathcal{G}$;

14   **if** $\sum_{k=1}^{temp} \min_{l \in W_{i,i+1}^{t,k}} B(l) \geq \overline{d^r}$ **then** allocate transmission resource for each kth path ($W_{i,i+1}^{t,k}$) using (2.28); success $\leftarrow$ true; break;

15   **end**

16   **if** *success = false* **then** break;

17 **end**

18 **if** *success = false* **then** $\mathcal{G}_v \leftarrow$ none;

19 **else return** $\mathcal{G}_v$;

If $\mathbb{A}(\mathcal{G}_v)$ is increased, we update the selected multicast topology with the new key-NFV node; If $\mathbb{A}(\mathcal{G}_v)$ is unchanged and $\mathbb{C}(\mathcal{G}_v)$ is reduced, we also update the selected multicast topology. If a path cannot accommodate all required VNFs (*i.e.*, $f_1, f_2, \ldots, f_{|\mathcal{V}|}$) after selecting a key-NFV node, we devise a corrective subroutine that instantiates the missing VNF instances on the closest NFV node from the multicast topology, and the corresponding transmission links are selected for traffic rerouting: Let P_t be the path from s to t in $\mathcal{G}_v$, $P_{t,f}$ be the union of the paths such that a missing function (f) is not hosted $\big($i.e.,$\{P_{t,f} = \cup_{t \in \mathcal{D}} P_t | f \text{ is not hosted}\}\big)$, and $P_{c,t,f}$ be longest common path before first branch in P. Correspondingly, for each missing VNF, we link the nearest applicable NFV node to $P_{c,t,f}$, place the missing VNF instance, and remove all unnecessary edges.

Until now, we construct a flexible multicast topology that connects the source to the destinations with all VNF instances traversed in order. We first start from the single-path scenario ($J = 1$) to check whether each path in $\mathcal{G}_v$ satisfies the link transmission rate requirement as per (2.13), and find an alternative path for each infeasible path. If a single-path solution is infeasible due to (2.13), the heuristic algorithm for the single-path case is extended to the multipath-enabled SFC embedding problem (with $J > 1$). Enabling multipath routing leads to several advantages: (1) It provides an alternative solution when the transmission rate requirement between two consecutively embedded VNF instances cannot be satisfied (i.e., when $B(l) < \overline{d^r}, l \in \mathcal{L}$), and (2) multipath routing is to reduce the overall link provisioning cost compared to the single-path case. For each path between two embedded VNF instances that fails to satisfy the transmission rate requirement $\overline{d^r}$, we add multipath routes incrementally, and search for a feasible solution up to the predefined J. The algorithm declares that the problem instance is infeasible when the number of multipath routes exceeds J.

For $J > 1$, we trigger a path splitting mechanism for each path between two embedded VNF instances for each S-D pair. Let $W_{i,i+1}^{t,k}$ be the kth path between two embedded VNFs (f_i, f_{i+1}) for destination t ($\in \mathcal{D}$), where the cardinality of all such possible paths is $K_{i,i+1}^t$. All candidate paths are first ranked in a descending order based on the amount of residual transmission resources. Then, we sequentially select from the candidate paths, such that the summation of all chosen paths' residual transmission rate meets the required transmission rate $\overline{d^r}$. Assuming the current number of trees is j, the transmission rates allocated on the kth path ($W_{i,i+1}^{t,k}$) is then calculated as

$$
\mathbb{R}(W_{i,i+1}^{t,k}) = \frac{B_{\min}^k \overline{d^r}}{\sum_{k=1}^{\min\{j,K_{i,i+1}^t\}} B_{\min}^k}, t \in \mathcal{D}, i \in \mathcal{S}_0^{|\mathcal{V}|}, k \in \mathcal{S}_1^{\min\{j,K_{i,i+1}^t\}}, r \in \mathcal{R}
$$

$$(2.28)$$

where $B_{\min}^k$ is the amount of minimum residual transmission resources for path $W_{i,i+1}^{t,k}$, i.e., $B_{\min}^k = \min_{l \in W_{i,i+1}^{t,k}} B(l)$. Here, the multipath extension method essentially computes a link-disjoint multipath configuration from a single-path route. Therefore, the proposed multipath extension is necessarily prone to the so-called path diminution problem, in which not all E2E multipath-enabled configurations can be devised from a single-path discovery [16, 17].

2.1.4.2 Multi-Service Scenario

For multiple SFC embedding, our strategy is to selectively prioritize the network services, contributing to the network throughput with least provisioning cost, to efficiently utilize the network resources. Here, we consider a static scenario, i.e., service requests are available a priori. Three principles serving as criteria to prioritize each service for embedding are identified. The first principle is to rank the given services based on the achieved network throughput, which is defined in (2.21). A service achieving high throughput is prioritized to be embedded on the substrate network, as it contributes more to the aggregate network throughput than a lower ranked service. However, ranking a service should also take into account the provisioning cost. It is impossible to obtain the exact service provisioning cost prior to the traffic routing and VNF placement process, as the problem is NP-hard. However, the relative positions of source and destinations provide implications on the cost incurred to provide the service. Given a service with highly distributive destinations, the provisioning cost is large, as more physical links and multicast replication points are needed to connect the destinations. Moreover, the distance from the source to destinations is in proportion to service provisioning cost as a long routing path with a large number of VNF instances is needed to establish a multicast topology.

Considering the distances between destinations and from source to destinations, we define a *distribution level*, denoted by g^r, as the product of two components. The first component is A^r/A, where A is the area of the smallest convex polygon that spans all nodes in the network, and A^r is the area of the smallest convex polygon that spans all destinations of service r. The ratio A^r/A provides an estimate of how dense a set of destinations for one service is in a given network area. Note that existing algorithms to determine the convex hull of a set of points and to calculate the area of an arbitrary shape are available [18]. The second component is q^r/q, where q^r is the distance from source to the center point of the set of destinations in service r, and q is the largest distance between two arbitrary nodes in the substrate network. The center point of the set of destinations in one service plays a role as a representer for all destinations in that service. The ratio q^r/q measures how far is the source from the destinations. The distribution level metric g^r is expressed as

$$g^r = \frac{A^r q^r}{Aq}, \quad r \in \mathcal{R}. \tag{2.29}$$

A larger value of g^r implies a higher distribution level, where the source is positioned further away from its destinations and the destinations are more distributive in the whole network coverage area. A largely distributive network service consumes more network resources, resulting in a high provisioning cost. Therefore, a service with a lower value of g^r has a higher priority to be embedded in order to preserve the substrate network resources. Although the parameters in (2.29) can be calculated with regard to the hop-count (e.g., via computing the shortest path) which is a more representative metric than the Euclidean distance, it is exhaustive to do so.

Next, we introduce a third metric named *size* to incorporate both (2.21) and (2.29) as

$$U^r = R^r(1 - g^r), \quad r \in \mathcal{R} \tag{2.30}$$

where the goal is to prioritize a service with higher throughput subject to a correction factor of $1 - g^r$ for how distributive (costly) it is.

To summarize, we calculate the throughput, the distribution level, and the size for each service request using (2.21), (2.29), and (2.30). Then, service requests are sorted according to their sizes in a descending order, and embedded using Algorithm 1.

2.1.5 Performance Evaluation

In this section, simulation results are presented to demonstrate the effectiveness of our proposed solutions to the joint multicast routing and VNF placement problems for the single and multiple service scenarios, with both single-path and multipath routing strategies enabled. The considered substrate network in our simulation is a mesh-topology based network [19], with $|\mathcal{N}| = 100$ and $|\mathcal{L}| = 684$, as shown in Fig. 2.4. We randomly choose 25 vertices as NFV nodes in the mesh network, and the transmission rate of physical link l and the computing rate of NFV node n are uniformly distributed between 0.5 and 2 million packet/s (Mpacket/s), i.e., $B(l), C(n) \sim \mathcal{U}(0.5, 2)$ Mpacket/s. To solve the formulated MILP problems, we use the Gurobi solver with the branch and bound algorithm, where the weighting coefficients are set as $\alpha = 0.6$, $\beta = 0.4$. The processing rate requirement of the VNFs are set to be linearly proportional to the incoming data rate, i.e., $C(f^r) = \overline{d^r}$ [20].

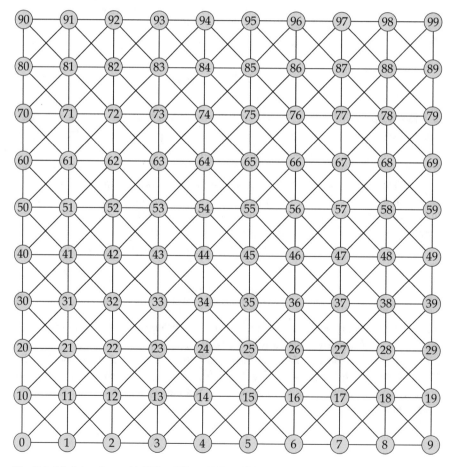

Fig. 2.4 Mesh topology with $|\mathcal{N}| = 100$ and $|\mathcal{L}| = 684$

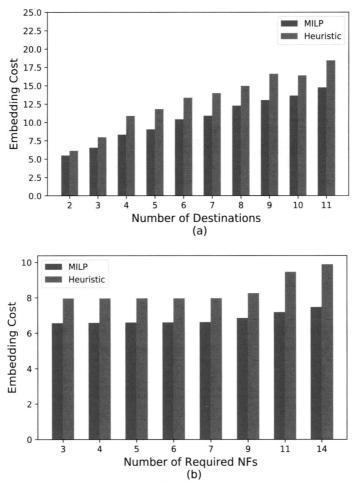

Fig. 2.5 Embedding cost with respect to (**a**) the number of destinations and (**b**) the number of required VNFs

First, we conduct a comparison between the optimal solution of the single-service single-path MILP formulation and the solution of the heuristic. We generate random service requests where the numbers of VNFs and destinations are varied from 3 to 14 and 2 to 11, respectively. The data rate of the generated service requests are set to $\overline{d^r} = 0.2$ Mpacket/s. The total provisioning cost obtained from both optimal and heuristic solutions are shown in Fig. 2.5, as the number of destinations $|\mathcal{D}|$ or VNFs $|\mathcal{V}|$ increases. It can be seen that the total provisioning cost increases with $|\mathcal{D}|$ or $|\mathcal{V}|$. As $|\mathcal{D}|$ increases, the costs obtained from both optimal and heuristic solutions grow with nearly a constant gap. Adding destinations incurs a higher cost than adding VNF instances, since additional physical links and VNF instances are required for each destination.

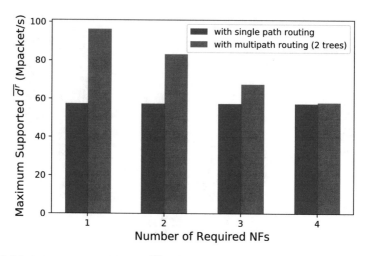

Fig. 2.6 Maximum supported data rate ($\overline{d^r}$) for both single-path and multipath routing scenarios using the proposed optimal formulation

Figure 2.6 shows the advantage of multipath routing over single-path routing using the proposed optimal formulation. We depict the maximum supported data rate $\overline{d^r}$, with which the VNF embedding is feasible. Compared to the single-path routing case, multipath routing ($J = 2$) always supports higher or equal data rates. With an increase of the number of required VNFs, the maximum supported data rate decreases, and converges to the single-path scenario; the processing cost becomes more significant and the number of candidate NFV nodes and paths decrease.

Next, we demonstrate the effectiveness of the proposed heuristic admission mechanism for the multi-service scenario discussed in Sect. 2.1.4.2. First, we divide the mesh topology into four access network regions and one core network region as indicated in Fig. 2.7. Three scenarios with different available processing and transmission rates on the NFV nodes and physical links are considered in the scale of Mpacket/s, as listed in Table 2.1. Each service randomly originates from one access network region, traverses the core network, and terminates in one of the other three access network regions. For each network scenario, to simulate network congestion, 35 multicast service requests are randomly generated and submitted for embedding, where the data rate $\overline{d^r}$ of service r is randomly distributed between [1.5, 3.5] Mpacket/s, and the number of VNFs and destinations are randomly generated as $|\mathcal{V}^r| = \{3, 4\}$ and $|\mathcal{D}^r| = \{3, 4, 5\}$. For each scenario, the service generation and embedding are repeated 5 times to average out the effect of randomness. As a benchmark, we use a second strategy that randomly selects the service requests for embedding.

Figure 2.8 shows the aggregate throughput $\mathbb{R}$ as in (2.20) achieved by the random and heuristic admission solutions under three scenarios specified in Table 2.1, which increases as the available processing and transmission rates increase. As shown in Fig. 2.8, the aggregate throughput of the proposed heuristic solution exceeds the

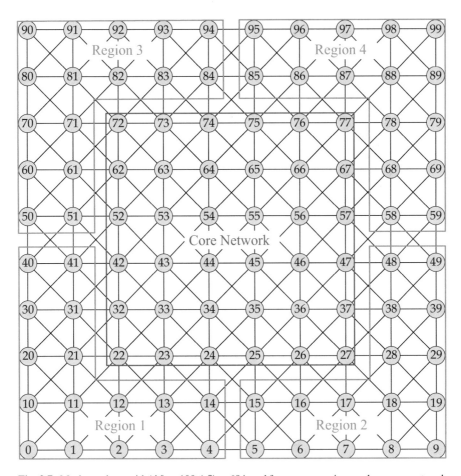

Fig. 2.7 Mesh topology with $|\mathcal{N}| = 100$, $|\mathcal{L}| = 684$, and four access regions and one core network

Table 2.1 Processing and transmission rates

Scenarios	Processing rate	Link transmission	NFV nodes
Scenario 1	$\mathcal{U}(3, 8)$ Mpacket/s	$\mathcal{U}(3, 8)$ Mpacket/s	47
Scenario 2	$\mathcal{U}(4, 9)$ Mpacket/s	$\mathcal{U}(4, 9)$ Mpacket/s	50
Scenario 3	$\mathcal{U}(5, 10)$ Mpacket/s	$\mathcal{U}(5, 10)$ Mpacket/s	53

aggregate throughput of the random admission by 15.63% on average over all the scenarios. This is because the size metric used in the heuristic solution ensures that the services with a larger throughput are embedded with a higher priority. Figure 2.9 shows the acceptance ratio of the total 35 service requests under different average data rates. The acceptance ratio of the heuristic solution exceeds the random solution over all source data rate levels by 4% on average.

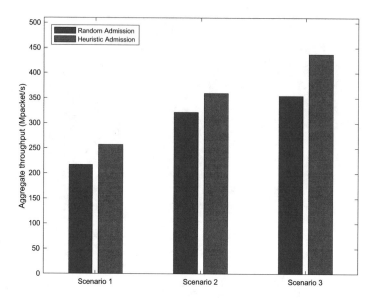

Fig. 2.8 Aggregate throughput comparison of the random admission and the proposed heuristic admission

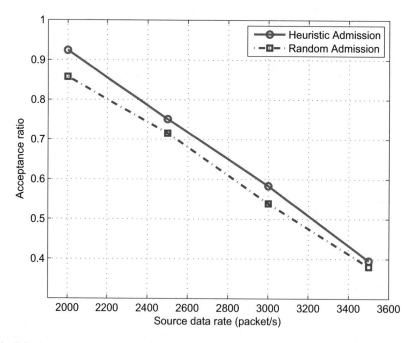

Fig. 2.9 Acceptance ratio comparison of the random admission and the proposed heuristic admission

2.2 E2E Delay Modeling for Embedded SFCs

E2E packet delay modeling of a delay-sensitive service flow traversing an embedded SFC is essential to evaluate the embedding performance. Existing research works investigate how to achieve optimal SFC embedding on the core network by minimizing the deployment, operation, and delay violation cost with network resource constraints and flow conservation constraints. E2E delay of forwarding each traffic flow is calculated as a summation of packet transmission delays on each physical link, without considering packet processing delay due to CPU processing on NFV nodes [21–23]. However, each packet from different traffic flows passing through an NFV node usually requires different amounts of time for CPU processing on the NFV node and for packet transmission on the outgoing link in sequence [10, 11]. Depending on the type of VNF each flow traverses, packets from some traffic flows are with large headers and thus bottleneck on CPU processing, whereas packets of other flows having large packet payload sizes consume more transmission time. Moreover, the packet arrival process of a flow at an NFV node correlates with packet processing and transmission at its preceding NFV nodes, making E2E delay analysis challenging. Therefore, how to develop an accurate analytical model to evaluate the delay that each packet of a traffic flow experiences when passing through an embedded SFC, including packet queueing delay, packet processing delay on NFV nodes, and packet transmission delay on links, is a challenging research issue. For SFC embedding, there is a fundamental tradeoff between satisfying E2E delay and reducing the SFC provisioning cost. Some SFCs are likely to be embedded on a common network path with multiple VNF instances hosted on one NFV node to improve resource utilization and reduce the deployment and operation cost for SFC provisioning. On the other hand, sharing a set of physical network resources with other traffic flows on NFV nodes and transmission links may increase the delay of one individual flow. Therefore, modeling E2E packet delay of each flow is of paramount importance to achieve delay-aware SFC embedding. As traffic flows traversing each NFV node demonstrate discrepant dominant (or bottleneck) resource consumption on either CPU or link transmission resources, how to slice the two resources among traffic flows to guarantee the fairness for resource sharing and maintain high resource utilization, referred to as *bi-resource slicing*, needs investigation and is correlated with E2E delay performance.

In this section, we first present dominant-resource generalized processor sharing (DR-GPS) [10] as the bi-resource slicing scheme among traffic flows sharing resources at each NFV node. The DR-GPS balances the tradeoff between service isolation and high resource utilization, and maintains dominant-resource fairness [12] among flows. Then, we model packet delay of a flow passing through NFV nodes and forwarding devices (e.g., physical links, network switches) of an embedded SFC in the following two steps:

1. Based on DR-GPS, we establish a tandem queueing model to extract the process of each flow going through CPU processing and link transmission at the first NFV node. To remove the rate coupling effect among flows, we derive average

processing rate as an approximation on instantaneous processing rate for each flow with the consideration of resource multiplexing. An M/D/1 queueing model is then developed for each decoupled processing queue to determine the average packet processing delay. Based on the analysis of packet departure process from each decoupled processing queue, the decoupled transmission rate is derived for each flow;

2. We analyze packet arrival process and then remove rate coupling effect of each flow traversing every subsequent NFV node of its embedded SFC. To eliminate the dependence of packet processing and transmission between consecutive NFV nodes, the arrival process of each flow at a subsequent NFV node is approximated as a Poisson process and an M/D/1 queueing model is employed to calculate the average delay for packet processing. It is proved that the average packet queueing delay for CPU processing calculated according to the approximated M/D/1 queueing model is an improved upper bound over that upon a G/D/1 queueing model. Therefore, an approximated M/D/1 queueing network is established to evaluate the average E2E packet delay of each flow traversing the entire SFC.

2.2.1 Service Function Chaining

As discussed in Sect. 1.1.3, logically, an aggregated data traffic flow of certain service type from a 5G wireless network is required to pass through a sequence of VNFs in a 5G core network to fulfill the E2E service delivery, as shown in Fig. 1.2. The sequential VNFs and the virtual links connecting them constitute an SFC. Over the physical core network, each VNF is instantiated and operated on an NFV node and each virtual link is represented by a sequence of physical transmission links and forwarding devices. This entire process is called *SFC embedding* or *service function chaining*. During the embedding/chaining process, both computing resources on servers and transmission resources on links are partitioned and reserved for different SFCs. We consider a set, I, of traffic flows passing through different SFCs over a same embedded network path. Figure 2.10 shows an example of two traffic flows, $i, j \in I$, traversing two SFCs, respectively, over a common embedded physical path and sharing same physical resources. Specifically, flow i from a DNS request service passes through the first NFV node N_1 operating the firewall function f_1 and transmission link L_0, which are then forwarded by n_1 network switches $\{R_1, \ldots, R_k, \ldots, R_{n_1}\}$ and n_1 transmission links $\{L_1, \ldots, L_k, \ldots, L_{n_1}\}$ before reaching the second NFV node N_2 operating DNS function f_2. At the same time, flow j traverses the same network path but passes through transcoding function f_3 on N_1 and then firewall function f_4 on N_2. After passing through N_2, traffic flows i and j are then forwarded by n_2 switches and n_2 links sequentially before reaching to the destination node in the core network. For a general case, a set, I, of traffic flows share a common embedded network path, with m NFV nodes denoted by N_y ($y = 1, 2, \ldots, m$) and n_y pairs of switches and transmission links forwarding traffic between NFV nodes N_y and N_{y+1} before arriving the destination.

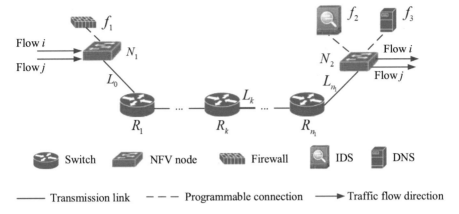

Fig. 2.10 SFC embedding over a same physical network path

With NFV and SDN, different VNFs can be flexibly placed at different server locations with optimized routing paths for traffic forwarding. Therefore, the overall computing and communication resource utilization is improved with reduced CapEx and OpEx. When a traffic flow traverses an NFV node, each packet of the flow consumes CPU time for packet processing and is then carried over the output link with certain transmission rate [10]. We assume the total amount of CPU resources are infinitely divisible on each network server [10, 11] and are shared among traffic flows passing through, and the transmission resources over links are also shared.

2.2.2 Resource Consumption Profile

As mentioned in Sect. 2.2.1, end user traffic of a same service type is aggregated as a traffic flow at the edge of the wireless network domain (e.g., traffic aggregation point), which is then forwarded through a core network by traversing a sequence of VNFs to fulfill certain service requirements. As illustrated in Fig. 2.10, packet arrivals of traffic flow i at the first NFV node, N_1, are modeled as a Poisson process with arrival rate λ_i. We define *time profile* for flow i traversing N_1 as a two-dimension vector $[\tau_{i,1}, \tau_{i,2}]$, which indicates that each packet from flow i requires $\tau_{i,1}$ amount of time for CPU processing and $\tau_{i,2}$ amount of time for link transmission if all processing resources on N_1 and transmission resources on L_0 are allocated to flow i [10]. Correspondingly, the rate vector $[C_{i,1}, C_{i,2}]$ (in packet per second) for flow i is the reciprocal of the time profile, $C_{i,1} = \frac{1}{\tau_{i,1}}$ and $C_{i,2} = \frac{1}{\tau_{i,2}}$. Traffic flows with different packet formats often have discrepant time profile when traversing an NFV node. For example, when passing through a firewall function, traffic flows with small packets but a large header size, such as DNS request packets, are more CPU demanding, whereas other traffic flows, e.g., video streaming traffic, with a large packet size consume more packet forwarding time than CPU processing. Therefore, traffic flows always demonstrate a more critical time consumption on either CPU processing or link transmission, which is referred to as *dominant resource*. Given a set of maximum available CPU resources on an NFV node and transmission resources its outgoing link, time profile of different traffic flows passing through NFV node can be discrepant. The available CPU resources on N_1 and transmission resources on L_0 are shared among flows. Suppose flow i is allocated packet processing rate $c_{i,1}$ out of the maximum processing rate $C_{i,1}$, and packet transmission rate $c_{i,2}$ out of the maximum transmission rate $C_{i,2}$. Note that when the CPU time is shared among multiple flows, there can be some overhead time when the CPU cores switch among flows for processing different tasks. Existing studies indicate that the switching overhead time only becomes comparable when the traffic load is saturated with a high CPU utilization percentage (i.e., CPU cores are frequently interrupted for switching tasks) [24, 25]. Considering only non-saturation traffic, we assume that the allocated processing rate $c_{i,1}$ for flow i ($\in I$) varies linearly with its occupied fraction of CPU time (i.e., the useful fraction of CPU usage). Thus, we denote the fraction of CPU resources allocated to flow i by $h_{i,1} = \frac{c_{i,1}}{C_{i,1}}$ and the fraction of link bandwidth resources by $h_{i,2} = \frac{c_{i,2}}{C_{i,2}}$.

2.2.3 Dominant-Resource Generalized Processor Sharing

When different traffic flows arrive at a common NFV node, we study how CPU and link transmission resources are sliced among the traffic flows, also called *bi-resource slicing*, to achieve high resource utilization and, at the same time, maintain a (weighted) fair allocation among the services. Since traffic flows from different services can have discrepant dominant resource types, the bi-resource slicing becomes more challenging than single resource slicing, in terms of balancing between high resource utilization and fair allocation for both resource types with QoS isolation.

The *generalized processor sharing (GPS)* discipline is a benchmark fluid-flow (i.e., infinitely divisible resources) based single resource sharing model [26, 27]. Each service flow, say flow i ($\in I$), at a common GPS server (e.g., a network switch) is assigned a positive priority value, φ_i, for transmission resource allocation. The GPS server guarantees that the allocated transmission rate g_i for flow i satisfies

$$g_i \geq \frac{\varphi_i}{\sum\limits_{i \in I} \varphi_i} G \tag{2.31}$$

where G is the maximum service rate of the GPS server. Note that the inequality sign in (2.31) holds when some flows in I do not have packets to transmit, and thus more resources can be allocated among any backlogged flows. Therefore, GPS has the properties of achieving both service isolation and a high resource multiplexing gain among flows.

However, if we directly apply GPS to the bi-resource context (i.e., bi-resource GPS), it is difficult to simultaneously maintain fair allocation for both CPU time and transmission rate, and achieve high system performance. Consider flow i and flow j, with time profiles $[\tau_{i,1}, \tau_{i,2}]$ and $[\tau_{j,1}, \tau_{j,2}]$ respectively, traversing NFV node N_1, with the same service priority for fair resource sharing. Assume $\tau_{i,1} > \tau_{i,2}$ and $\tau_{j,1} < \tau_{j,2}$. If we apply the bi-resource GPS, both the maximum processing and transmission rates are equally divided for the two flows. Consequently, the performance of both flows traversing N_1 is not maximized: For flow i, due to imbalanced time profiles, the allocated link transmission rate $c_{i,2}$ is larger than the processing rate $c_{i,1}$, leading to resource wastage on link transmission; The situation reverses for flow j, where packets are accumulated/queued for transmissions, leading to the queueing delay increase. Therefore, to improve the system performance, a basic principle [10] is that the fractions, $h_{i,1}$ and $h_{i,2}$, of CPU and transmission resources allocated to any flow i ($\in I$) should be in proportion to its time profile, i.e., $\frac{h_{i,1}}{h_{i,2}} = \frac{\tau_{i,1}}{\tau_{i,2}}$, to guarantee the allocated processing rate and the transmission rate are equalized, i.e., $c_{i,1} = c_{i,2}$. In such a way, queueing delay before packet transmissions of each flow is minimized. However, based on this principle, if we apply GPS only on one of the two resources (i.e., single-resource GPS with equalized processing and transmission rates), the allocation of the other type of resources among traffic flows continues to be unbalanced due to the time profile discrepancy of among flows.

To balance the trade-off between high service performance and fair allocation of both types of resources among flows, we employ a *dominant-resource generalized processor sharing* scheme [10] for the bi-resource allocation. The DR-GPS combines dominant resource fairness (DRF) [12] with GPS, where the fractions of allocated dominant resources among different backlogged flows are equalized based on service priority, and the other type of resources are allocated to ensure the rates of packet processing and transmission are equalized for each flow (the basic principle is applied). When some flows do not have packets backlogged for processing, their allocated resources are redistributed among other backlogged flows (if any). As there are a finite combinations of flows forming a backlogged flow set out of I, we denote B ($\in I$) as one of the combinations for a backlogged flow set. Each backlogged flow has a dominant resource consumption in either packet processing or transmission. The DR-GPS discipline is mathematically formulated in (P1) when the set, I ($|I| \geq 1$), of traffic flows traverse N_1, where $|\cdot|$ is the set cardinality.

$$(P1) : \max\{h_{1,1}, ..., h_{i,1}, ..., h_{j,1}, ..., h_{|B|,1}\}$$

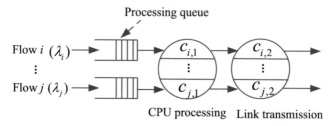

Fig. 2.11 Queueing modeling for traffic flows traversing NFV node N_1

$$
\text{s.t.} \begin{cases}
\displaystyle\sum_{i \in B} h_{i,1} \leq 1 & \text{(2.32a)} \\[2ex]
\displaystyle\sum_{i \in B} h_{i,2} \leq 1 & \text{(2.32b)} \\[2ex]
h_{i,1} = \dfrac{\tau_{i,1}}{\tau_{i,2}} h_{i,2} & \text{(2.32c)} \\[2ex]
\dfrac{h_{i,d}}{w_i} = \dfrac{h_{j,d}}{w_j}, \quad \forall i, j \in B & \text{(2.32d)} \\[2ex]
h_{i,1}, h_{i,2}, w_i, w_j \in [0, 1]. & \text{(2.32e)}
\end{cases}
$$

In (P1), w_i and w_j are the resource allocation weights, representing service priorities for flow i and flow j, respectively, and $h_{i,d}$ is the fraction of occupied dominant resources of flow i, which is either $h_{i,1}$ or $h_{i,2}$. Constraint (2.32c) guarantees $c_{i,1} = c_{i,2}$; Constraint (2.32d) equalizes the fractions of allocated dominant resources among the backlogged flows. Problem (P1) is a linear programming problem and can be solved efficiently to obtain the optimal solutions of $h_{i,1}$ and $h_{i,2}$ for any flow i.

The DR-GPS has properties of (1) QoS isolation by guaranteeing a minimum service rate as in (2.31) for each flow, and (2) work conservation by fully utilizing at least one of the two types of resources in serving the backlogged flows [10, 28, 29]. Although the queueing delay for packet transmissions is reduced by employing the DR-GPS scheme, the total packet delay[1] for each flow traversing a common NFV node should be evaluated. With GPS [26, 27], the process of multiple traffic flows passing through an NFV node, for example, node N_1, is extracted as a tandem queueing model, shown in Fig. 2.11. The total delay is a summation of packet queueing delay for CPU processing, processing delay, and transmission delay. In the following, we develop an analytical model to evaluate the total delay for each

---

[1] Total delay is defined as the duration from the time that a packet of one traffic flow reaches the processing queue of an NFV node to the instant that it is transmitted over an output link of the node.

packet of a traffic flow traversing the first NFV node (e.g., node N_1), upon which the E2E delay for packets passing through an embedded SFC is analyzed.

2.2.4 E2E Delay Modeling

In this section, we first analyze the total delay for each packet of a traffic flow traversing the first NFV node, and then extend the analytical delay model to evaluate the E2E packet delay for traffic lows passing through a sequence of NFV nodes of an embedded SFC.

2.2.4.1 Processing and Transmission Rate Decoupling

The main difficulty of packet delay analysis comes from the dependencies of both processing and transmission rates of each flow on the backlog status of other flows multiplexed at a common NFV node. For the case of two traffic flows, when one flow has an empty processing queue, its processing and transmission resources are utilized by the other backlogged flow to exploit the traffic multiplexing gain. Therefore, the processing (transmission) rate of each flow varies between two deterministic values, depending on the backlog status of the other flow. For a general case of I multiplexed flows which contain a set, B, of backlogged flows excluding the tagged flow i, the processing rate $c_{i,1}$ of flow i is determined by solving (P1). We further denote B as B_r where $r = 1, 2, \ldots, \binom{|I|}{|B|}$, representing one of $\binom{|I|}{|B|}$ combinations of $|B|$ backlogged flows. Thus, $c_{i,1}$ changes with B_r. This correlation of queueing status among flows causes the variation of the flow processing rate among discrete deterministic values, posing challenges on the total packet delay analysis.

To remove the coupling of instantaneous processing rates among different flows, we first determine the average processing rate $\mu_{i,1}$ for flow i, with the consideration of resource multiplexing among flows, i.e., non-empty probabilities of processing queues, through a set of high-order nonlinear equations, given by

$$\begin{cases} \mu_{i,1} = \sum_{|B|=0}^{|I|-1} \sum_{r=1}^{M} \prod_{l \in B_r} \varrho_{l,1} \prod_{k \in \overline{B_r}} (1 - \varrho_{k,1}) c_{i,1} \\ \varrho_{i,1} = \dfrac{\lambda_i}{\mu_{i,1}}, \quad \forall i \in I. \end{cases} \tag{2.33}$$

In (2.33), $M = \binom{|I|-1}{|B|}$, $\overline{B_r} = I \setminus \{i \bigcup B_r\}$, and $\varrho_{i,1}$ is the non-empty probability of processing queue for flow i at N_1. Given packet arrival rate for any flow in I, (2.33) has $2|I|$ equations with $2|I|$ variables and can be solved numerically for the set of average processing rates of each flow. For analysis tractability, we use the

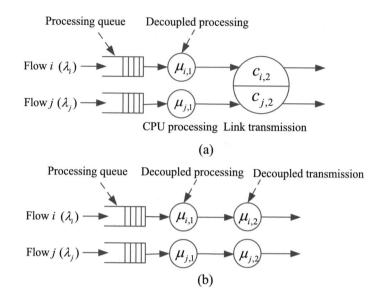

Fig. 2.12 A queueing model for (**a**) decoupled packet processing and (**b**) decoupled packet processing and transmission

average processing rate, $\mu_{i,1}$, as an approximation for the instantaneous processing rate $c_{i,1}$ to transform the original correlated processing system to a number of $|I|$ uncorrelated processing queues. A case of two traffic flows at an NFV node is demonstrated in Fig. 2.12. With the decoupled deterministic processing rates, packet processing for each flow is modeled as an M/D/1 queueing system, upon which we further calculate both the decoupled packet processing delay and the total delay for packet processing.

In Fig. 2.12a, the instantaneous link transmission rates among the traffic flows are also correlated. To remove the transmission rate correlation for delay analysis tractability, we determine the packet arrival process of each flow at the output link of the first NFV node (i.e., packet departure process after CPU processing) in the following.

2.2.4.2 Delay Modeling at the First NFV Node

The packet departure process from each decoupled processing at N_1 is analyzed. Taking flow i in Fig. 2.12 as an example, the inter-departure time between two successively departed packets from the decoupled processing is calculated. As indicated in [30] and [31], for an M/D/1 queueing system, the steady-state queue occupancy distribution seen by a departing packet is the same as that seen by an arriving packet due to the Poisson packet arrival process. Therefore, a departing packet from the decoupled processing sees the same empty probability of the processing queue as an arriving packet. Let random variable Y_i be the inter-departure time of successive packets of flow i departing from the decoupled processing at N_1. If the l th departing packet sees a nonempty queue, then $Y_i = T_i$, where $T_i = \frac{1}{\mu_{i,1}}$ is the decoupled processing time for a packet of flow i; If the departing packet sees an empty queue upon its departure, $Y_i = X_i + T_i$, where random variable X_i denotes the duration from the time of the l th packet departure of flow i to the time of $(l+1)$ th packet arrival. Because of the memoryless property of a Poisson arrival process, X_i has the same exponential distribution as packet inter-arrival time with parameter λ_i. Therefore, the probability density function (PDF) of Y_i, $\xi_{Y_i}(t)$, can be calculated as

$$\xi_{Y_i}(t) = \left(1 - \varrho_{i,1}\right) \xi_{(X_i + T_i)}(t) + \varrho_{i,1}\xi_{T_i}(t) \tag{2.34}$$

where $\varrho_{i,1} = \frac{\lambda_i}{\mu_{i,1}}$. Since X_i and T_i are independent random variables, the PDF of $X_i + T_i$ is the convolution of the PDFs of X_i and T_i. Thus, (2.34) is further derived as

$$
\begin{aligned}
\xi_{Y_i}(t) &= \left(1 - \frac{\lambda_i}{\mu_{i,1}}\right)\left[\xi_{X_i}(t) \circledast \xi_{T_i}(t)\right] + \frac{\lambda_i}{\mu_{i,1}}\xi_{T_i}(t) \\
&= \left(1 - \frac{\lambda_i}{\mu_{i,1}}\right)\left[\lambda_i e^{-\lambda_i t}u(t) \circledast \delta(t - T_i)\right] \\
&\quad + \frac{\lambda_i}{\mu_{i,1}}\delta(t - T_i) \\
&= \frac{\lambda_i\left(\mu_{i,1} - \lambda_i\right)}{\mu_{i,1}}e^{-\lambda_i(t - T_i)}u(t - T_i) + \frac{\lambda_i}{\mu_{i,1}}\delta(t - T_i)
\end{aligned}
\tag{2.35}
$$

where $u(t)$ is the unit step function, $\delta(t)$ is the Dirac delta function, and $\circledast$ is the convolution operator. From (2.35), the cumulative distribution function (CDF) of Y_i is given by

$$F_{Y_i}(t) = \left[1 - \left(1 - \frac{\lambda_i}{\mu_{i,1}}\right)e^{-\lambda_i(t - T_i)}\right]u(t - T_i). \tag{2.36}$$

Based on (2.35) and (2.36), both mean and variance of Y_i can be calculated as

$$E[Y_i] = \frac{1}{\lambda_i} \quad \text{and} \quad D[Y_i] = \frac{1}{\lambda_i^2} - \frac{1}{\mu_{i,1}^2}. \tag{2.37}$$

From (2.36) and (2.37), we observe that, when λ_i is small, the departure process, delayed by the service time T_i, approaches the Poisson arrival process with parameter λ_i; when λ_i is increased to approach $\mu_{i,1}$, the departure process approaches the deterministic process with rate $\mu_{i,1}$.

Since the packet departure rate from each decoupled processing is the same as packet arrival rate at each processing queue, the decoupled transmission rate $\mu_{i,2}$ for flow i is the same as $\mu_{i,1}$, by solving the set of equations in (2.33). At this point, we have a complete decoupled queueing model of both packet processing and packet transmission for flow i traversing the first NFV node N_1, shown in Fig. 2.12b. The average total packet delay for flow i is determined by

$$D_{i,1} = \frac{1}{\mu_{i,1}} + \frac{\lambda_i}{2\mu_{i,1}^2(1 - \varrho_{i,1})} + \frac{1}{\mu_{i,2}}. \tag{2.38}$$

Before modeling the delay for flows traversing the second NFV node, N_2, we analyze the packet departure process of each flow for link transmission at N_1. Similar to the analysis of packet departure process from the decoupled processing for flow i at N_1, we set time 0 as the instant when the l th packet departs from the processing queue and reaches the transmitting queue for immediate packet transmission. Since we have $\mu_{i,1} = \mu_{i,2}$, T_i also indicates packet transmission time, i.e., $T_i = \frac{1}{\mu_{i,2}}$. Let Z_i denote packet inter-departure time for flow i passing though link transmission. If the l th departing packet from the processing queue sees a nonempty queue, we have $Z_i = T_i$; otherwise, the following two cases apply, as illustrated in Fig. 2.13:

Case 1—If the $(l + 1)$ th packet's arrival time at the processing queue is greater than the l th packet's transmission time at the transmitting queue, i.e, $X_i > T_i$, we have

$$Z_i = \zeta_1 + 2\zeta_2 = (X_i - T_i) + 2T_i = X_i + T_i \tag{2.39}$$

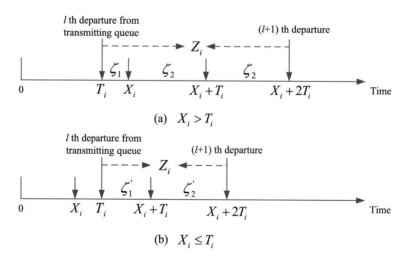

Fig. 2.13 A composition of Z_i under different cases

where ζ_1 indicates the duration from the instant that the l th packet departs from the transmission queue till the instant that the $(l + 1)$ th packet arrives at the processing queue, and $\zeta_2 = T_i$;

Case 2—If the $(l + 1)$ th packet arrives at the processing queue while the l th packet is still at the transmission queue, i.e., $X_i \leq T_i$, we have

$$Z_i = \zeta_1' + \zeta_2' = [T_i - (T_i - X_i)] + T_i = X_i + T_i \qquad (2.40)$$

where ζ_1' denotes the remaining processing time on the $(l + 1)$ th packet at the processing queue after the l th packet departs from transmission queue, and $\zeta_2' = T_i$.

As a result, the PDF of Z_i is given by

$$
\begin{aligned}
\xi_{Z_i}(t) = \left(1 - \varrho_{i,1}\right) & \left[P\{X_i \leq T_i\}\xi_{(X_i + T_i)}(t) \right. \\
& \left. + P\{X_i > T_i\}\xi_{(X_i + T_i)}(t)\right] + \varrho_{i,1}\xi_{T_i}(t) \\
= \left(1 - \varrho_{i,1}\right) & \xi_{(X_i + T_i)}(t) + \varrho_{i,1}\xi_{T_i}(t).
\end{aligned}
\qquad (2.41)
$$

Comparing (2.34) and (2.41), we conclude that Z_i and Y_i have exactly the same probability distribution, and thus any order of statistics (e.g. expectation, variance).

2.2.4.3 Delay over the Virtual Link Between NFV Nodes

The packet departure process of each flow from the link transmission at the first NFV node N_1 is derived. Before reaching the second NFV node N_2, the flows may be forwarded by a sequence of network switches and over physical transmission links. The transmission rates allocated to flow i from these forwarding devices are the same as the transmission rates $\mu_{i,2}$ from N_1 for maximizing the transmission resource utilization [32], with which the queueing delays on switches and links are minimized and negligible for each flow. The total packet transmission delay for flow i traversing n_1 switches and n_1 links before reaching N_2 is given by

$$D_{i,1}^{(f)} = \frac{2n_1}{\mu_{i,2}}. \qquad (2.42)$$

2.2.4.4 Average E2E Delay

Proposition 1 *A Poisson packet flow traverses a sequence of k ($k = 1, 2, 3, \ldots$) network servers, each with deterministic service capacity $Y^{(k)}$. If we have $Y^{(q)} \geq Y^{(q-1)}$, $\forall q \in [2, k]$, the departure process of the traffic flow coming out of the k th ($k \geq 2$) server remains the same as the departure process from the first server.*

The proof of Proposition 1 is given in Appendix A.

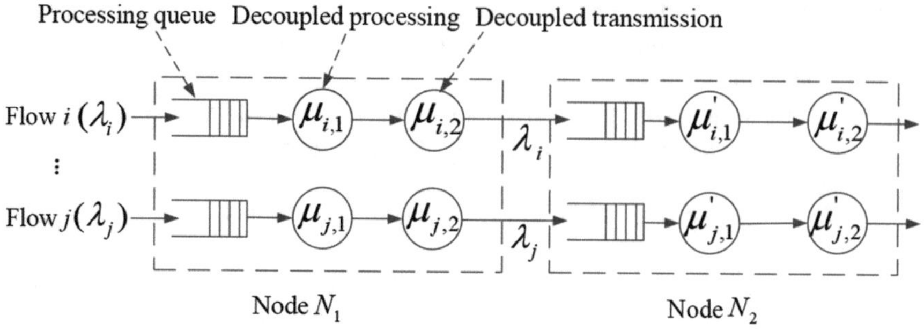

Processing queue Decoupled processing Decoupled transmission

Fig. 2.14 A decoupled queueing model for traffic flows traversing the first and second NFV nodes in sequence

According to Proposition 1, the arrival process at the subsequent NFV node N_2 following N_1 is the same as the traffic departure process from N_1. Based on (2.37) and (2.41), the arrival rate for flow i at N_2 is λ_i. Thus, the same methodology is applied as in (2.33) to obtain a set of decoupled processing and transmission rates $\mu'_{i,1}$ and $\mu'_{i,2}$ for flow i at N_2, as shown in Fig. 2.14, by taking into consideration the time profiles of traffic flows traversing the new VNF(s) at N_2 and instantaneous processing and transmission rates $c'_{i,1}$ and $c'_{i,2}$ allocated to flow i. The main difference in delay modeling for flow i at N_2 and at N_1 is that the packet arrival process for flow i is a general process with average arrival rate λ_i. The process has the inter-arrival time Z_i with the same CDF, expectation and variance as those of Y_i in (2.36) and (2.37). We can model the CPU processing for flow i at N_2 as a G/D/1 queueing process,[2] where the average packet queueing delay before processing is given by Bertsekas et al. [30]

$$W_{i,2} = \frac{\lambda_i \left(\frac{1}{\lambda_i^2} - \frac{1}{\mu_{i,1}^2} - \sigma_i^2 \right)}{2 \left(1 - \varrho'_{i,1} \right)} \leq \frac{\lambda_i \left(\frac{1}{\lambda_i^2} - \frac{1}{\mu_{i,1}^2} \right)}{2 \left(1 - \varrho'_{i,1} \right)}. \tag{2.43}$$

In (2.43), $\rho'_{i,1} = \frac{\lambda_i}{\mu'_{i,1}}$, T_e is the idle duration within inter-arrival time of successive packets of flow i at N_2, with variance σ_i^2.

Since the flow arrival process at N_2 correlates with the preceding decoupled processing rates at N_1, as indicated in (2.36), the G/D/1 queueing model is not accurate especially when λ_i becomes large [30]. Also, it is difficult to obtain the distribution of T_e to calculate σ_i^2 in (2.43). It is not accurate to use the upper bound in (2.43) to approximate $W_{i,2}$ when λ_i is small (the queueing system is lightly loaded), since the probability of an arriving packet at the processing queue of N_2 seeing an empty queue increases, and σ_i^2 becomes large.

---

[2] Note that we consider the case where $\mu'_{i,1} < \mu_{i,2}$, $\forall i \in I$; For the case of $\mu'_{i,1} \geq \mu_{i,2}$, there is no queueing delay for processing at N_2.

From (2.36) and (2.37), in the case of $\mu'_{i,1} < \mu_{i,2}$, Z_i is more likely to approach an exponentially distributed random variable than a deterministic value with varying λ_i under the condition of $\rho'_{i,1} < 1$. Therefore, to make the arrival process of each flow at the processing queue of N_2 independent of processing and transmission rates at N_1, we approximate packet arrival process of flow i at N_2 as a Poisson process with rate parameter λ_i, and establish an M/D/1 queueing model to represent the packet processing for flow i at N_2. Proposition 2 indicates that the average packet queueing delay $Q_{i,2}$ in the M/D/1 queueing system is an improved upper bound over that in the G/D/1 system in (2.43), especially when the input traffic is lightly loaded.

Proposition 2 *Given $\mu'_{i,1} < \mu_{i,2}$, $Q_{i,2}$ is an upper bound of $W_{i,2}$ when the processing queue for flow i at N_2 is both lightly-loaded and heavily-loaded.*

A detailed proof of Proposition 2 is provided in Appendix B.

According to the approximation on packet arrival process for flow i at N_2, average packet delay passing through N_2 is calculated, independently of the processing and transmission rates at N_1, given by

$$
D_{i,2} = \begin{cases} \dfrac{1}{\mu'_{i,1}} + \dfrac{\lambda_i}{2{\mu'_{i,1}}^2 \left(1 - \rho'_{i,1}\right)} + \dfrac{1}{\mu'_{i,2}}, & \mu'_{i,1} < \mu_{i,2} \\[2ex] \dfrac{1}{\mu'_{i,1}} + \dfrac{1}{\mu'_{i,2}}, & \mu'_{i,1} \geq \mu_{i,2}. \end{cases} \tag{2.44}
$$

Based on the same methodology of delay modeling for packets traversing N_2, the average packet delay for flow i passing through each subsequent NFV node (if any) is determined independently. Under the condition that the decoupled packet processing rate of flow i at one subsequent NFV node N_z ($z > 2$) is smaller than the decoupled packet transmission rate at its preceding NFV node N_{z-1}, an approximated M/D/1 queueing process is used to represent CPU processing for flow i at N_z, which is valid as the packet arrival process is more likely to approach a Poisson process with varying λ_i. In general, an approximated M/D/1 queueing network is established to calculate the average E2E delay for each packet of flow i traversing an embedded SFC with m NFV nodes. The E2E delay is the summation of average delay for packet processing on all NFV nodes and the total transmission delay on network switches and links along the flow forwarding path, given by

$$
D_i = \sum_{z=1}^{m} D_{i,z} + \sum_{z=1}^{m} D_{i,z}^{(f)}. \tag{2.45}
$$

In (2.45), $D_{i,z}$ is the average delay for a packet of flow i traversing zth NFV node of the embedded SFC, $D_{i,z}^{(f)}$ is the total packet transmission delay for flow i passing through n_z switches and n_z links before reaching NFV node N_{z+1}, and is calculated as in (2.42).

2.2.5 Performance Evaluation

Computer simulations are conducted to verify the accuracy of the proposed analytical model for evaluating the E2E delay of traffic flows passing through embedded SFCs. All simulations are carried out using the network simulator OMNeT++ [33]. Consider two equally weighted flows, i and j, traversing the embedded SFCs firewall (f_1) → DNS (f_3) and firewall (f_1) → IDS (f_2) respectively, over a common physical path in a 5G core network, and sharing a set of processing and transmission resources, as shown in Fig. 1.2. Flow i represents DNS request traffic, whereas flow j indicates a video-conferencing service. The packet arrival rate λ_i for flow i is set to 150 packet/s with packet size of 4000 bits [34]. The packet size for flow j is set to 16,000 bits, and we vary its arrival rate, λ_j, from 75 packet/s to 350 packet/s to reflect different traffic load conditions. The rate vector for each flow traversing an NFV node is tested over OpenStack [35], which is a resource virtualization platform installed on each NFV node. By injecting traffic flows with different packet sizes into different VNFs programmed on the OpenStack, we test maximum available packet processing and transmission rates for different flows. With DR-GPS, each flow is allocated a fraction of the maximum available processing and transmission rates, upon which packet-level simulation is conducted to evaluate packet delay of each flow traversing each NFV node. Table 2.2 summarizes the rate vectors for flows i and j traversing different VNFs. Other important simulation settings are also included.

2.2.5.1 Delay at the First NFV Node

First, the packet processing delay and packet queueing delay for each flow traversing the first NFV node N_1 are compared. It is demonstrated in Fig. 2.15a and b that both packet processing delay and packet queueing delay obtained based on rate decoupling are close to the simulation results with rate coupling. The delay for link transmission and packet queueing are also evaluated for both flows in Fig. 2.16. For transmission delay, the analytical results match the simulation results well, and

Table 2.2 Simulation parameters

	Traffic flows	
Parameters	Flow i	Flow j
Rate vector (Firewall)	[1000, 2000] packet/s	[750, 500] packet/s
Rate vector (DNS)	[1000, 1250] packet/s	–
Rate vector (IDS)	–	[800, 312.5] packet/s
n_1	20	20
n_2	25	25
Simulation time	1000 s	1000 s

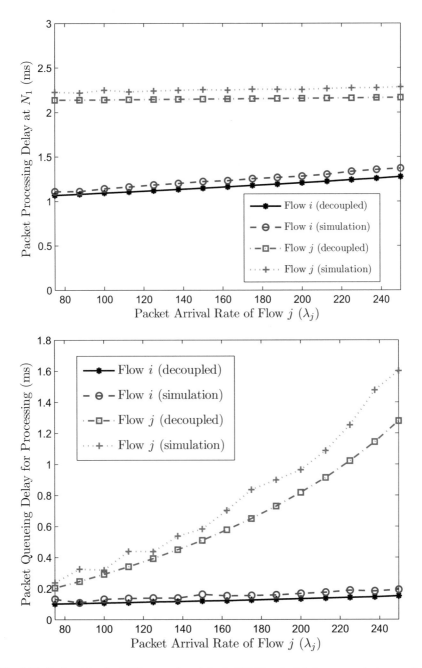

Fig. 2.15 Packet delay for CPU processing at the first NFV node; (**a**) Average packet processing delay (**b**) Average packet queueing delay

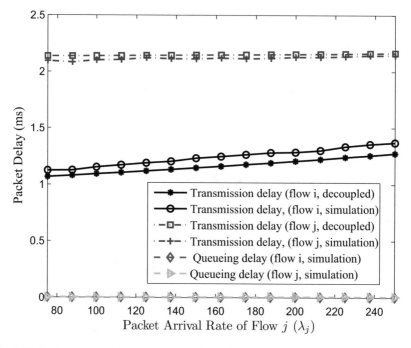

Fig. 2.16 Packet transmission and queueing delay at N_1

the packet queueing delay for link transmission is minimal and negligible, which verifies the accuracy of the proposed transmission rate decoupling.

2.2.5.2   Delay at Subsequent NFV Nodes

After traversing the first NFV node, packet processing delay for flow i and flow j at the subsequent NFV node, N_2, is evaluated in Fig. 2.17. As the analytical and simulation results closely match, it is verified that the processing rate decoupling at N_2 is accurate. Packet queueing delay at N_2 for CPU processing for both flows are demonstrated in Figs. 2.18 and 2.19. As λ_j increases, the packet queueing delay for flow i increases slightly since the resource multiplexing gain obtained by flow i from flow j becomes small. It is seen from Fig. 2.18 that the M/D/1 queueing model for packet processing at N_2 provides a tighter packet queueing delay upper bound than the G/D/1 queueing model for flow i. In comparison, it is observed in Fig. 2.19 that the G/D/1 upper bound for packet queueing delay of flow j is tighter than that of flow i, since the processing queue for flow i is lightly loaded with low queue nonempty probability. From Fig. 2.19, the G/D/1 queueing delay upper bound becomes tighter for flow j with the increase of λ_j but less accurate in a heavy traffic load condition. For both flows, an improved upper bound is calculated

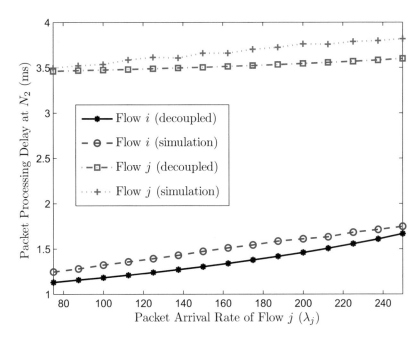

Fig. 2.17 Packet processing delay at N_2

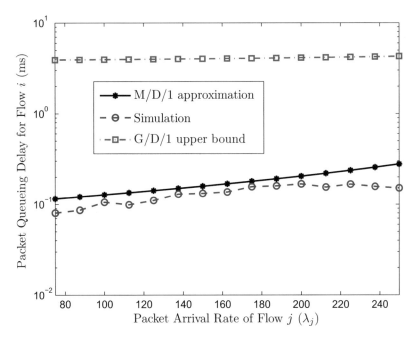

Fig. 2.18 Packet queueing delay of flow i at N_2

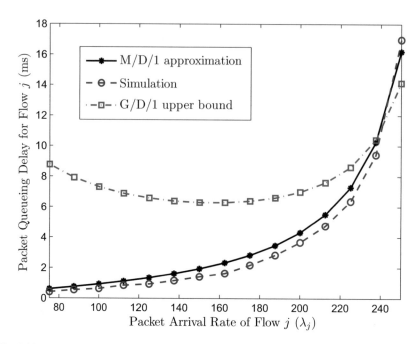

Fig. 2.19 Packet queueing delay of flow j at N_2

to approximate the packet queueing delay for CPU processing at N_2, based on the proposed M/D/1 queueing model.

For both flows, the packet transmission delay at N_2 is evaluated in Fig. 2.20. Similar to Fig. 2.16, it is seen that the decoupled transmission rates are close to the simulation results, and queueing delays for link transmission are negligible. The end-to-end delay for packets of each flow traversing the whole embedded physical network path is also evaluate in Fig. 2.21, which is a summation of packet processing and queueing delays for traversing all NFV nodes and packet transmission delays over all embedded physical links and switches. It is shown that the proposed analytical delay modeling provides an accurate evaluation for multiple traffic flows traversing embedded SFCs over a common physical network path. The proposed analytical modeling also promotes the realization of delay-aware SFC embedding.

2.3 Summary

In this chapter, we have studied joint traffic routing and VNF placement problems for both single and multiple multicast service scenarios over an SDN/NFV-enabled physical substrate network. For the single-service scenario, an optimization problem is formulated to minimize function and link provisioning cost under the physical

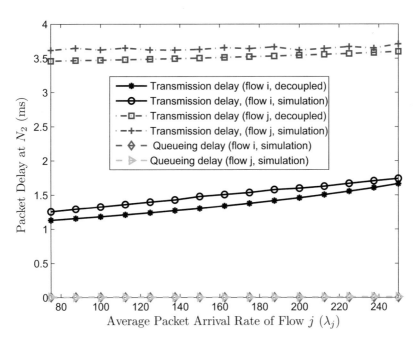

Fig. 2.20 Delay for packet transmissions at N_2

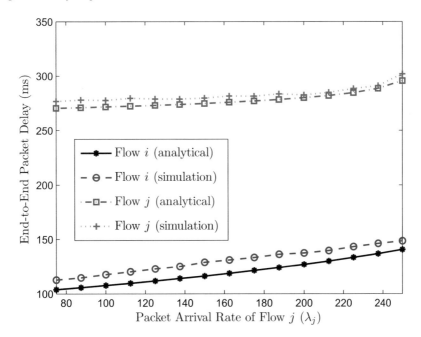

Fig. 2.21 E2E packet delay for both flows

resource and flow conservation constraints; For the multi-service case, we present an optimization framework that jointly deals with multiple service requests, such that the aggregate network throughput is maximized with minimal function and link provisioning costs. To realize delay-aware SFC embedding, we have further established an analytical model to evaluate E2E packet delay for multiple traffic flows traversing their embedded SFCs. With DR-GPS, both CPU and transmission resources are allocated among different flows at each NFV node to achieve dominant-resource fair allocation with high resource utilization. Through processing and transmission rate decoupling, we establish an approximated M/D/1 queueing network to evaluate the average E2E delay for packets from different traffic flows traversing the embedded SFCs. Extensive simulation results demonstrate the accuracy and effectiveness of our proposed virtual network topology design frameworks and analytical E2E packet delay model for achieving delay-aware SFC embedding.

References

1. M. Zhao, B. Jia, M. Wu, H. Yu, Y. Xu, Software defined network-enabled multicast for multi-party video conferencing systems, in *Proc. IEEE ICC* (2014), pp. 1729–1735
2. C. Qiao, H. Yu, J. Fan, W. Zhong, X. Cao, X. Gao, Y. Zhao, Z. Ye, Virtual network mapping for multicast services with max-min fairness of reliability. J. Opt. Commun. Netw. **7**(9), 942–951 (2015)
3. J. Blendin, J. Ruckert, T. Volk, D. Hausheer, Adaptive software defined multicast, in *Proc. IEEE NetSoft* (2015), pp. 1–9
4. O. Alhussein, P.T. Do, J. Li, Q. Ye, W. Shi, W. Zhuang, X. Shen, X. Li, J. Rao, Joint VNF placement and multicast traffic routing in 5G core networks, in *Proc. IEEE GLOBECOM* (2018), pp. 1–6
5. S.Q. Zhang, Q. Zhang, H. Bannazadeh, A. Leon-Garcia, Routing algorithms for network function virtualization enabled multicast topology on SDN. IEEE Trans. Netw. Serv. Manag. **12**(4), 580–594 (2015)
6. S.Q. Zhang, A. Tizghadam, B. Park, H. Bannazadeh, A. Leon-Garcia, Joint NFV placement and routing for multicast service on SDN, in *Proc. IEEE NOMS* (2016), pp. 333–341
7. M. Zeng, W. Fang, Z. Zhu, Orchestrating tree-type VNF forwarding graphs in inter-DC elastic optical networks. J. Lightw. Technol. **34**(14), 3330–3341 (2016)
8. Z. Xu, W. Liang, M. Huang, M. Jia, S. Guo, A. Galis, Approximation and online algorithms for NFV-enabled multicasting in SDNs, in *Proc. IEEE ICDCS* (2017), pp. 625–634
9. J.-J. Kuo, S.-H. Shen, M.-H. Yang, D.-N. Yang, M.-J. Tsai, W.-T. Chen, Service overlay forest embedding for software-defined cloud networks, in *Proc. IEEE ICDCS* (2017), pp. 720–730
10. W. Wang, B. Liang, B. Li, Multi-resource generalized processor sharing for packet processing, in *Proc. ACM IWQoS' 13* (2013, June), pp. 1–10
11. A. Ghodsi, V. Sekar, M. Zaharia, I. Stoica, Multi-resource fair queueing for packet processing. ACM SIGCOMM Comput. Commun. Rev. **42**(4), 1–12 (2012)
12. A. Ghodsi, M. Zaharia, B. Hindman, A. Konwinski, S. Shenker, I. Stoica, Dominant resource fairness: Fair allocation of multiple resource types, in *Proc. USENIX NSDI* (2011), pp. 323–336
13. S. Mehraghdam, M. Keller, H. Karl, Specifying and placing chains of virtual network functions, in *Proc. IEEE CloudNet* (2014), pp. 7–13
14. M. Chowdhury, M.R. Rahman, R. Boutaba, ViNEYard: Virtual network embedding algorithms with coordinated node and link mapping. IEEE/ACM Trans. Netw. **20**(1), 206–219 (2012)

15. O. Alhussein, P.T. Do, Q. Ye, J. Li, W. Shi, W. Zhuang, X. Shen, X. Li, J. Rao, A virtual network customization framework for multicast services in NFV-enabled core networks. IEEE J. Sel. Areas Commun. **38**(6), 1025–1039 (2020)
16. A.M. Abbas, B.N. Jain, Mitigating path diminution in disjoint multipath routing for mobile ad hoc networks. Int. J. Ad Hoc Ubiquitous Comput. **1**(3), 137–146 (2006)
17. A.M. Abbas, B.N. Jain, Path diminution in node-disjoint multipath routing for mobile ad hoc networks is unavoidable with single route discovery. Int. J. Ad Hoc Ubiquitous Comput. **5**(1), 7–21 (2010)
18. K.R. Anderson, A reevaluation of an efficient algorithm for determining the convex hull of a finite planar set. Inf. Process. Lett. **7**(1), 53–55 (1978)
19. N. Zhang, Y. Liu, H. Farmanbar, T. Chang, M. Hong, Z. Luo, Network slicing for service-oriented networks under resource constraints. IEEE J. Sel. Areas Commun. **35**(11), 2512–2521 (2017)
20. M.T. Beck, J.F. Botero, Coordinated allocation of service function chains, in *Proc. IEEE Globecom* (2015), pp. 1–6
21. L. Wang, Z. Lu, X. Wen, R. Knopp, R. Gupta, Joint optimization of service function chaining and resource allocation in network function virtualization. IEEE Access **4**, 8084–8094 (2016)
22. F. Bari, S.R. Chowdhury, R. Ahmed, R. Boutaba, O.C.M.B. Duarte, Orchestrating virtualized network functions. IEEE Trans. Netw. Serv. Manag. **13**(4), 725–739 (2016)
23. S. Ayoubi, C. Assi, K. Shaban, L. Narayanan, MINTED: Multicast virtual network embedding in cloud data centers with delay constraints. IEEE Trans. Commun. **63**(4), 1291–1305 (2015)
24. X. Li, J. Rao, H. Zhang, A. Callard, Network slicing with elastic SFC, in *Proc. IEEE VTC' 17* (2017, Sept.), pp. 1–5
25. N. Egi et al., Understanding the packet processing capability of multi-core servers. Intel Tech. Rep. (2009)
26. A.K. Parekh, R.G. Gallager, A generalized processor sharing approach to flow control in integrated services networks: The single-node case. IEEE/ACM Trans. Netw. **1**(3), 344–357 (1993)
27. Z.-L. Zhang, D. Towsley, J. Kurose, Statistical analysis of the generalized processor sharing scheduling discipline. IEEE J. Sel. Areas Commun. **13**(6), 1071–1080 (1995)
28. D.C. Parkes, A.D. Procaccia, N. Shah, Beyond dominant resource fairness: Extensions, limitations, and indivisibilities. ACM Trans. Econ. Comput. **3**(1), 3 (2015)
29. A. Gutman, N. Nisan, Fair allocation without trade, in *Proc. ACM AAMAS' 12* (2012, June), pp. 719–728
30. D.P. Bertsekas, R.G. Gallager, P. Humblet, *Data Networks*, vol. 2 (Prentice-Hall, Englewood Cliffs, NJ, 1987)
31. L. Kleinrock, *Queueing Systems, Volume 1: Theory* (Wiley, New York, 1976)
32. S. Larsen, P. Sarangam, R. Huggahalli, S. Kulkarni, Architectural breakdown of end-to-end latency in a TCP/IP network. Int. J. Parallel Program. **37**(6), 556–571 (2009)
33. OMNeT++ 5.0. [Online]. Available: http://www.omnetpp.org/omnetpp
34. A. Liska, G. Stowe, *DNS Security: Defending the Domain Name System* (Syngress, Cambridge, 2016)
35. Openstack (Release Pike). [Online]. Available: https://www.openstack.org. Accessed Dec. 2017

Chapter 3
Dynamic Resource Slicing for Service Provisioning

3.1 Bi-resource Slicing for 5G Core Networks

In a 5G core network, when multiple traffic flows of different service types traverse an NFV node, both CPU and link transmission resources need to be sliced properly and allocated to each flow to achieve high resource utilization with fair resource sharing among flows. As discussed in Sect. 2.2.3, with GPS, each traffic flow, say flow $x(\in J)$, multiplexing at a common a network element (e.g., a network server/switch/link), is guaranteed a minimum service rate, $\frac{\psi_x}{\sum\limits_{x \in J} \psi_x} R$, if all flows have backlogged packets to be transmitted, where ψ_x is a positive weighting factor reflecting the resource sharing priority of flow x and R is the maximum packet service rate of the network element. When some of the flows have empty transmission queues, their allocated transmission/processing rates are re-distributed among the remaining backlogged flows to exploit the traffic multiplexing gain. The GPS has properties of achieving QoS isolation among flows with improved utilization of a single resource type.

However, when the GPS is directly applied to bi-resource slicing at an NFV node for multiple flows, it is difficult to achieve high performance in both CPU processing and link transmission and to maintain a fair resource slicing among flows, since traffic flows have discrepant dominant resource consumption profiles. Suppose two equally-weighted flows x and y are considered traversing firewall function F_1 at V_1 as shown in Fig. 3.1. The two flows have resource profiles $[t_{x,1}, t_{x,2}]$ and $[t_{y,1}, t_{y,2}]$ respectively, with dominant resource consumption on different resource types, i.e., $t_{x,1} > t_{x,2}$ and $t_{y,1} < t_{y,2}$. As discussed in Sect. 2.2.3, to achieve a low packet delay and maintain a fair allocation on both types of resources, we employ a DR-GPS scheme [1], which combines dominant resource fairness [2] with GPS, for bi-resource slicing among multiple flows traversing an NFV node. With DR-GPS, the fractions of dominant resources allocated to each backlogged flow are equalized, and the non-dominant resources are allocated in proportion to the

© The Author(s), under exclusive license to Springer Nature Switzerland AG 2021
Q. Ye, W. Zhuang, *Intelligent Resource Management for Network Slicing in 5G and Beyond*, Wireless Networks, https://doi.org/10.1007/978-3-030-88666-0_3

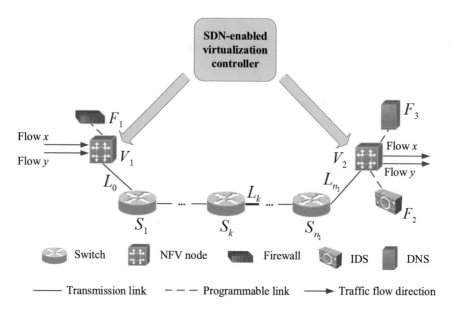

resource consumption profiles of each flow to guarantee equalized packet processing and transmission rates. When some of the flows have no packets to be transmitted, the allocated resources are re-distributed among other backlogged flows (one of the properties of GPS). In the example of two backlogged flows x and y at the NFV node V_1, with BR-GPS, the fraction of CPU resources allocated to flow x equals the fraction of link transmission resources allocated to flow y, i.e., $f_{x,1} = f_{y,2}$, and the allocation on the other resource type follows guarantees $r_{x,1} = r_{x,2}$ and $r_{y,1} = r_{y,2}$.

Since each flow demands more on its dominant resource type, the DR-GPS equalizes the shares of dominant resources for a fair allocation among backlogged flows. More importantly, by equalizing the allocated processing and transmission rates, the DR-GPS minimizes the packet queueing delay at the outgoing transmission link of an NFV node. With the GPS properties, the DR-GPS achieves service isolation by guaranteeing each backlogged flow minimum fractions of CPU and transmission resources, and achieves high resource utilization by traffic multiplexing [1]. The proposed bi-resource slicing scheme is also compared with the following two resource sharing under consideration:

1. Bi-resource GPS: When we have two backlogged flows, the fractions of CPU and transmission resources sliced to flow x and flow y are equalized (applying GPS on both resource types), i.e., $f_{x,i} = f_{y,i} = \frac{1}{2}$ ($i = 1, 2$), where $f_{x,i} = \frac{r_{x,i}}{R_{x,i}}$ and $f_{y,i} = \frac{r_{y,i}}{R_{y,i}}$, and $r_{x,i}$ and $r_{y,i}$ denote the allocated packet processing or packet transmission rates, respectively, to flow x and flow y. However, due to the discrepant resource consumption profiles, the equal partitioning on both CPU and transmission resources between the flows cause imbalanced packet processing

and transmission rates. For flow x, the link transmission resources are overly provisioned as we have $r_{x,1} < r_{x,2}$; For flow y, since the allocated processing rate is greater than the transmission rate, i.e., $r_{y,1} > r_{y,2}$, packets are accumulated for link transmission, resulting in a long packet queueing delay;

2. Single-resource GPS with equalized service rates (applying GPS on one resource type): To reduce the total delay for packets from both backlogged flows traversing the NFV node, a basic principle is to allocate the fractions of CPU and transmission resources for each flow in proportion to its resource consumption profiles, i.e., $\frac{f_{x,1}}{f_{x,2}} = \frac{t_{x,1}}{t_{x,2}}$. In such a way, the allocated processing and transmission rates are equalized. However, if we apply GPS on one type of resources while the other type of resources are allocated accordingly to achieve equalized packet processing and transmission rates, the resource utilization on the other type is imbalanced between the two flows due to the discrepant resource consumption profiles.

3.1.1 Performance Comparison

For the comparison of different bi-resource slicing schemes, Fig. 3.2 shows the resource shares for flows x and y based on single-resource GPS and DR-GPS, respectively. The DR-GPS equalizes the fractions of allocated dominant resources for flows x and y. The CPU processing rate is also equalized with the link transmission rate for each flow. Therefore, the DR-GPS achieves a dominant-resource fair allocation with high resource utilization between the flows. In contrast, applying the single-resource GPS with equalized processing and transmission rates causes imbalanced transmission resource utilization between the flows. In Fig. 3.3, we compare packet queueing delays of flows x and y for link transmission at V_1 by

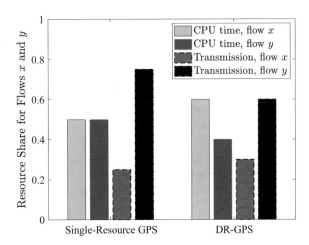

Fig. 3.2 Fractions of allocated resources to flows x and y under different resource slicing schemes

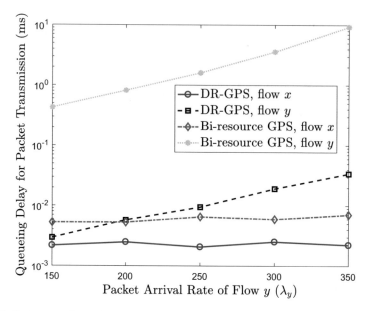

Fig. 3.3 Packet queueing and transmission delay under different resource slicing schemes

employing DR-GPS and bi-resource GPS schemes. Although the bi-resource GPS achieves fair allocation on both CPU and transmission resources between the two flows, the amount of allocated transmission resources is over-provisioned for flow x and is underestimated for flow y, due to discrepant resource consumption profiles of different flows. Therefore, packet queueing delay of flow y for link transmission under the bi-resource GPS is much longer than that under the DR-GPS where packet queueing delays of both flows at the transmission link are minimized.

3.2 Dynamic Radio Resource Slicing for 5G Wireless Networks

Radio resource slicing for a 5G wireless network requires a network-level resource partitioning, where the radio resources on different BSs are aggregated as a resource pool and are partitioned into a number of resource slices for each BS with service quality guarantee; Resources on each BS are further scheduled among end users/devices for wireless transmissions in the network operation stage. Since users/devices from each service provider (SP) are dispersed over network cells, the entire radio resources are logically sliced for different SPs, but are physically partitioned among BSs.

Existing studies mainly focus on service-level resource orchestration [3–5], where radio spectrum resources at each BS are pre-configured according to specified policies and are sliced among different groups of service users/devices under the coverage of the BS. However, the network-level resource sharing and slicing among BSs needs investigation for maximizing the resource utilization, where NFV is leveraged to enable radio resource abstraction and aggregation for the purpose of centrally controlling the resources on wireless BSs. Some studies address spectrum-level resource sharing among long-term evolution (LTE) BSs without explicitly differentiating traffic statistics and QoS descriptions among diversified services (i.e., mobile broadband data service and M2M service) [6, 7]. Therefore, to satisfy differentiated service requirements, traffic modeling for supported services should be considered in resource slicing. Also, the impact of network dynamics on the slicing performance is to be investigated. In this section, we present a dynamic radio resource slicing framework for a two-tier 5G wireless network to facilitate spectrum sharing among BSs with differentiated QoS isolation.

3.2.1 System Model

As illustrated in Sect. 1.1.2, we consider a two-tier downlink 5G wireless access network, where an MBS, denoted by B_m, is deployed for a wide area coverage, and a set of SBSs, $\mathcal{B} = \{B_k, k = 1, 2, \ldots, n\}$ (n is the number of small cells) are randomly distributed underlaying the coverage of the macro-cell, to support machine-type devices (MTDs) and mobile users (MUs), as shown in Fig. 3.4. There are two categories of MTDs and MUs, where a category I device within the macro-

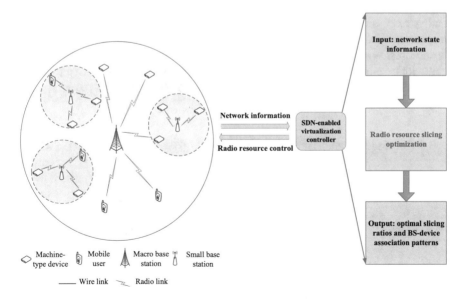

Fig. 3.4 A two-tier 5G wireless access network with radio resource slicing

cell coverage is connected to the MBS for wireless transmissions but is outside the coverages of all small cells, and a category II device stays within the coverages of both the macro-cell and one of the small cells, which chooses to be associated with either the MBS or the SBS depending on the network conditions, e.g., cell loads, channel conditions and network utility. We denote the set of category I MUs and MTDs along with their set cardinalities as $\mathcal{N}_u$ and N_u, and $\mathcal{N}_a$ and N_a, respectively, while the sets of category II MUs and MTDs and their set cardinalities in the coverage of B_k ($k = 1, 2, \ldots, n$) are denoted by $\mathcal{M}_k$ and M_k and $\mathcal{N}_k$ and N_k, among which some of the devices can be associated with the MBS. There are a number of link-layer packet transmission queues at each BS for downlink transmissions from the BS to an connected device. Link-layer packetized traffic is used to model data traffic and M2M traffic arrivals for explicit QoS characterization [8]. Data packets, each with constant size of L_d bits, intended to be transmitted to an MU arrive at a transmission queue in the MBS periodically with rate λ_d packet/s, whereas machine-type packets for an MTD arrive at the MBS or an SBS in an event-driven manner with a much lower packet arrival rate and a smaller packet size [9]. As suggested in [10, 11], machine-type packet arrivals at each transmission queue of a BS are modeled as a Poisson process with average rate of λ_a packet/s and constant packet size L_a bits. Due to the randomness of machine-type traffic arrivals, the QoS is guaranteed in a statistical way by applying the effective bandwidth theory [8].

Two portions of radio resources, W_m and W_s, are pre-configured on the MBS and each SBS, respectively, which are mutually orthogonal to avoid inter-network-tier interference. As each SBS has a small communication coverage with low transmit power, the radio resources W_s are reused among all SBSs with controlled inter-cell interference to exploit high spectrum utilization. Transmit power for B_m and B_k ($k = 1, 2, \ldots, n$), denoted by P_m and P_k ($k = 1, 2, \ldots, n$), are pre-configured and remain constant during each resource slicing period. The slicing of radio resources is updated when the network traffic load fluctuates [6]. A cell traffic load is described as a combination of number of devices currently served by the network cell with certain traffic load on each device. Based on Shannon capacity formula, the spectrum efficiency at device i from B_m and B_k ($k = 1, 2, \ldots, n$) are given, respectively, by

$$r_{i,m} = \log_2 \left(1 + \frac{P_m G_{i,m}}{\sigma^2} \right), \, i \in \{\mathcal{N}_u, \mathcal{N}_a, \mathcal{N}_k, \mathcal{M}_k\}, k = \{1, 2, \ldots, n\} \qquad (3.1)$$

and

$$r_{i,k} = \log_2 \left(1 + \frac{P_k G_{i,k}}{\sum\limits_{j=\{1,2,\ldots,n\}, j \neq k} P_j G_{i,j} + \sigma^2} \right), \, i \in \mathcal{N}_k \cup \mathcal{M}_k, k = \{1, 2, \ldots, n\}$$

$$(3.2)$$

where $G_{i,m}$ and $G_{i,k}$ are squared average channel gains between B_m and B_k and device i, σ^2 denotes average background noise power. Inter-small-cell interference is included in (3.2). In (3.1) and (3.2), $r_{i,m}$ or $r_{i,k}$ is termed as *spectrum efficiency* which is measured in a large timescale to capture long-term wireless channel conditions (e.g., path loss) [12]. The downlink achievable rate at each MTD or MU depends on BS-device association patterns and downlink radio resources allocated from B_m or S_k. Note that the sliced radio resources among BSs are updated in a large timescale based on statistics of network state information [12]. Therefore, the received signal-to-noise (SNR) and signal-to-interference-plus-noise (SINR) in (3.1) and (3.2) are average SNR and SINR over each resource slicing period [6], where the fast channel fading effects are averaged out.

3.2.2 Dynamic Radio Resource Slicing Framework

For a two-tier 5G wireless network supporting differentiated services, the utilization of pre-configured radio resources on BSs can be inefficient due to imbalanced user/device distribution and varying network traffic load [13]. In addition, a massive number of MTDs occupying the network resources for data/control information transmission can possibly lead to QoS violation on existing MUs if resources are not efficiently allocated between the data service and M2M service [14, 15]. Thus, radio resource slicing is required to balance the trade-off between (1) facilitating resource sharing among BSs, and (2) achieving QoS isolation among differentiated services. On one hand, resources are sliced and reserved to M2M service and data service for differentiated QoS satisfaction; On the other hand, the amount of resources for each slice should be dynamically adjusted by adapting to network conditions for improved resource utilization. Therefore, a dynamic radio resource slicing framework is developed with the consideration of both network-level (e.g., network traffic load and dynamics) and service-level information (e.g., differentiated traffic characteristics and service requirements). Moreover, as MUs are with low-to-moderate mobility and MTDs are in most cases stationary, resource slices are updated in a large timescale based on statistical network state information exchange between the control module and the BSs.

With NFV and SDN, all transmission resources on BSs are centrally controlled, the amount of which is denoted by W_v ($W_v = W_m + W_s$). As shown in Fig. 3.4, based on the network state information (e.g., number of end devices, traffic load on each device, wireless channel conditions), the central controller needs to specify how the aggregated radio resources are sliced among BSs to achieve maximal resource utilization and, at the same time, guarantee QoS isolation between M2M and data services. We define β_m and β_s (with $\beta_m + \beta_s = 1$) as *slicing ratios*, indicating the shares of radio resources (out of W_v) sliced to the MBS and SBSs, respectively. To determine the optimal set of slicing ratios, $\{\beta_m^*, \beta_s^*\}$, we formulate a comprehensive optimization framework to maximize the aggregate network utility under the constraints of satisfying the differentiated QoS requirements for both types of services, by taking into consideration the network conditions and service-level characteristics (see details in Sect. 3.2.3). The slicing ratios are also adjusted in response to the network traffic load dynamics to improve the overall resource utilization. The operation procedure of the radio resource slicing optimization includes the following three steps:

Step 1 Through signaling exchange between the SDN-enabled NFV controller and the BSs, the controller periodically collects the updated network state information, including the total number of MTDs and MUs from the M2M service and the broadband data service within the coverage area of each BS, traffic load statistics λ_a and λ_d, and wireless channel conditions between the BSs and end devices (users) considering the inter-small-cell interference.

Step 2 Based on the updated network information, the controller runs the radio resource slicing optimization to maximize the overall communication resource utilization under differentiated QoS constraints for both M2M and broadband data services, upon which the optimal slicing ratios and the optimal BS-MTD (MU) association patterns, are obtained.

Step 3 Repeat *Step 2* to update the optimal radio resource slicing ratios and BS-MTD (MU) association patterns, when the network traffic load varies, to achieve consistently maximum network-level radio resource utilization.

3.2.3 Problem Formulation

In the radio resource slicing framework, the main issue is how to determine the optimal slicing ratios, $\{\beta_m^*, \beta_s^*\}$, on BSs to (1) maximize the aggregate network resource utilization, and (2) meet differentiated QoS requirements for M2M and data services. The following logarithmic functions are considered as utility functions for an end device (MU or MTD) associated with B_m and B_k, respectively:

$$\log\left(c_{i,m}\right) = \log\left(W_v\beta_m f_{i,m}r_{i,m}\right), i \in \mathcal{N}_u \cup \mathcal{N}_a \cup \mathcal{N}_k \cup \mathcal{M}_k, k = \{1, 2, \dots, n\} \tag{3.3}$$

$$\log\left(c_{i,k}\right) = \log\left(W_v\beta_s f_{i,k}r_{i,k}\right), i \in \mathcal{N}_k \cup \mathcal{M}_k, k = \{1, 2, \dots, n\}. \tag{3.4}$$

In (3.3) and (3.4), $c_{i,m}$ denotes the downlink average achievable rate at device i connected to B_m, and $c_{i,k}$ the average achievable rate for device i (category II) from B_k; $f_{i,m}$ is the fraction of radio resources (out of $W_v\beta_m$) allocated to device i from B_m, and $f_{i,k}$ the fraction of radio resources (out of $W_v\beta_s$) allocated to device i (category II) from B_k. The logarithmic utility function is a concave function with diminishing marginal utility, which facilitates network load balancing and resource allocation fairness among end devices [5, 6, 13].

Note that the average achievable rates, $c_{i,m}$ and $c_{i,k}$, are assumed constant during each resource slicing period. Therefore, the throughput requirement for each MU can be satisfied deterministically as long as enough resources are allocated for each downlink data transmission pair. However, due to stochastic packet arrivals from the event-driven M2M service, the downlink packet transmission delay from B_m or B_k to an MTD should be guaranteed statistically for maximal resource utilization. The delay is the duration from the instant a machine-type packet arrives at a BS transmission queue to the instant the intended MTD receives the packet. We apply the effective bandwidth theory [16] in calculating the minimum transmission rate $c^{(min)}$, for each MTD to probabilistically guarantee an upper-bounded packet transmission delay violation probability ε, given by

$$c^{(min)} = -\frac{L_a \log \varepsilon}{D_{max} \log\left(1 - \frac{\log \varepsilon}{\lambda_a D_{max}}\right)}. \tag{3.5}$$

A detailed derivation of $c^{(min)}$ can be found in [17]. Then, we formulate an aggregate network utility maximization problem to optimize the radio resource slicing ratios on BSs, with differentiated QoS guarantee for both M2M and mobile broadband data services, as given in (P1):

$$\max_{\substack{\beta_m,\beta_s, \\ x_{i,j}, f_{i,j}}} \sum_{i\in\mathcal{N}_u\cup\mathcal{N}_a} \log(c_{i,m}) + \sum_{k=1}^{n} \sum_{i\in\mathcal{N}_k\cup\mathcal{M}_k} \sum_{j\in\{m,k\}} x_{i,j} \log(c_{i,j})$$

$$c_{i,m} \geq \lambda_d L_d, \; i \in \mathcal{N}_u \tag{3.6a}$$

$$c_{i,m} \geq c^{(min)}, \; i \in \mathcal{N}_a \tag{3.6b}$$

$$x_{i,j}\left[c_{i,j} - c^{(min)}\right] \geq 0, \; i \in \mathcal{N}_k, \; j \in \{m, k\} \tag{3.6c}$$

$$x_{i,j}\left[c_{i,j} - \lambda_d L_d\right] \geq 0, \; i \in \mathcal{M}_k, \; j \in \{m, k\} \tag{3.6d}$$

$$\sum_{j \in \{m,k\}} x_{i,j} = 1, \; i \in \mathcal{N}_k \cup \mathcal{M}_k \tag{3.6e}$$

s.t. $\Bigg\{$

$$x_{i,j} \in \{0, 1\}, \; i \in \mathcal{N}_k \cup \mathcal{M}_k, \; j \in \{m, k\} \tag{3.6f}$$

$$\sum_{i \in \mathcal{N}_u \cup \mathcal{N}_a} f_{i,m} + \sum_{k=1}^{n} \sum_{i \in \mathcal{N}_k \cup \mathcal{M}_k} x_{i,m} f_{i,m} = 1 \tag{3.6g}$$

$$\sum_{i \in \mathcal{N}_k \cup \mathcal{M}_k} x_{i,k} f_{i,k} = 1 \tag{3.6h}$$

$$f_{i,m} \in (0, 1), \; i \in \mathcal{N}_u \cup \mathcal{N}_a \cup \mathcal{N}_k \cup \mathcal{M}_k \tag{3.6i}$$

$$f_{i,k} \in (0, 1), \; i \in \mathcal{N}_k \cup \mathcal{M}_k \tag{3.6j}$$

$$\beta_m + \beta_s = 1 \tag{3.6k}$$

$$\beta_m, \beta_s \in [0, 1]. \tag{3.6l}$$

The objective function in (P1) is the aggregate network utility, the summation of utilities achieved by all devices. As a category I device is associated with B_m and a category II device chooses to connect to either B_m or B_k, a binary variable, $x_{i,j}$ ($i \in \mathcal{N}_k \cup \mathcal{M}_k, j \in \{m, k\}, k \in \{1, 2, \ldots, n\}$), is used to indicate the association pattern for category II device i with B_m or B_k. Device i is associated with B_m, if $x_{i,m} = 1$ and $x_{i,k} = 0$; Otherwise, it is associated with B_k, if $x_{i,m} = 0$ and $x_{i,k} = 1$.

In (P1), constraints (3.6a) and (3.6d) indicate that the downlink transmission rate $c_{i,j}$ for any MU i is greater than or equal to the periodic data packet arrival rate destined for the device at the base station. Constraints (3.6b) and (3.6c) ensure that the achievable rate for both category I and category II MTDs is not less than the effective bandwidth of the M2M traffic source. Constraint (3.6e) and (3.6f) indicate that a category II device is associated with either the MBS or its home SBS during each radio resource allocation period. Constraints (3.6g) and (3.6h) indicate the requirements on resource allocation to MTDs and MUs from different BSs. Therefore, by maximizing the aggregate network utility with QoS guarantee, the optimal slicing ratios β_m^* and β_s^*, BS-device association patterns $x_{i,j}^*$, and fractions of radio resources $f_{i,m}^*$ and $f_{i,k}^*$ allocated to MTDs and MUs from B_m and B_k are obtained.

As (P1) is a joint BS-device association and resource allocation problem, the fraction of radio resources allocated to each device are correlated with the BS-device association patterns and the ratios of sliced resources on BSs, which makes the problem difficult to solve. For tractability, given BS-device association patterns $x_{i,j}$ and resource slicing ratios $\{\beta_m, \beta_s\}$, we solve (P1) for the optimal fractions of radio resources $f_{i,m}^*$ and $f_{i,k}^*$ allocated to device i from B_m and B_k, as a function of $x_{i,j}$, to reduce the number of decision variables.

3.2.3.1   Optimizing Resource Allocation for MUs and MTDs

As discussed, we solve (P1) for $f_{i,j}$, as a function of $x_{i,j}$, which is then substituted into (P1) to reduce the number of decision variables. Specifically, given β_m, β_s, and $x_{i,j}$, the objective function of (P1) is expressed as a summation of $u_m^{(1)}(f_{i,m})$ and $\sum_{k=1}^{n} u_k^{(1)}(f_{i,k})$, where $u_m^{(1)}(f_{i,m})$ is a function of $f_{i,m}$, indicating the aggregate utility of MUs and MTDs associated with B_m, given by

$$u_m^{(1)}(f_{i,m}) = \sum_{i \in N_u \cup N_a} \log\left(W_v \beta_m f_{i,m} r_{i,m}\right) + \sum_{k=1}^{n} \sum_{i \in \mathcal{N}_k'} \log\left(W_v \beta_m f_{i,m} r_{i,m}\right) \qquad (3.7)$$

where $\mathcal{N}_k' = \{l \in N_k \cup M_k | x_{l,m} = 1\}$, $u_k^{(1)}(f_{i,k})$ is a function of $f_{i,k}$, indicating the aggregate utility of category II devices associating with S_k, given by

$$u_k^{(1)}(f_{i,k}) = \sum_{i \in \overline{\mathcal{N}_k'}} \log\left(W_v \beta_s f_{i,k} r_{i,k}\right) \qquad (3.8)$$

with $\overline{\mathcal{N}_k'} = \{l \in N_k \cup M_k | x_{l,k} = 1\}$. Thus, (P1) is expressed as (P1'):

$$\max_{f_{i,m}, f_{i,k}} u_m^{(1)}(f_{i,m}) + \sum_{k=1}^{n} u_k^{(1)}(f_{i,k})$$

$$\text{s.t.} \begin{cases} \sum\limits_{i \in N_u \cup N_a \cup \mathcal{N}_k'} f_{i,m} = 1 & (3.9a) \\[2ex] \sum\limits_{i \in \overline{\mathcal{N}_k'}} f_{i,k} = 1 & (3.9b) \\[2ex] f_{i,m} \in (0, 1),\ i \in N_u \cup N_a \cup \mathcal{N}_k' & (3.9c) \\[1ex] f_{i,k} \in (0, 1),\ i \in \overline{\mathcal{N}_k'}. & (3.9d) \end{cases}$$

In (P1'), $\{f_{i,m}\}$ and $\{f_{i,k}\}$ are two independent sets of decision variables with uncoupled constraints. Thus, (P1') is further decomposed to the following two subproblems:

$$(\text{S1P1'}) : \max_{f_{i,m}} u_m^{(1)}(f_{i,m})$$

$$\text{s.t.} \begin{cases} \sum_{i \in \mathcal{N}_u \cup \mathcal{N}_a \cup \mathcal{N}'_k} f_{i,m} = 1 & (3.10\text{a}) \\ \\ f_{i,m} \in (0, 1), \quad i \in \mathcal{N}_u \cup \mathcal{N}_a \cup \mathcal{N}'_k & (3.10\text{b}) \end{cases}$$

and

$$(\text{S2P1'}) : \max_{f_{i,k}} \sum_{k=1}^{n} u_k^{(1)}(f_{i,k})$$

$$\text{s.t.} \begin{cases} \sum_{i \in \mathcal{N}'_k} f_{i,k} = 1 & (3.11\text{a}) \\ \\ f_{i,k} \in (0, 1), \quad i \in \overline{\mathcal{N}'_k}. & (3.11\text{b}) \end{cases}$$

Proposition 3 *The solutions for (S1P1') and (S2P1') are*

$$f_{i,m}^* = \frac{1}{N_u + N_a + \sum\limits_{k=1}^{n} \sum\limits_{l \in \mathcal{N}_k \cup \mathcal{M}_k} x_{l,m}} \triangleq f_m^* \tag{3.12}$$

and

$$f_{i,k}^* = \frac{1}{\sum\limits_{l \in \mathcal{N}_k \cup \mathcal{M}_k} x_{l,k}} \triangleq f_k^*. \tag{3.13}$$

The proof of Proposition 3 is provided in Appendix C. Proposition 3 indicates that the optimal fractions of radio resources allocated to MUs and MTDs from an associated BS are equalized.

By substituting f_m^* and f_k^* into (P1), the problem is reformulated with the reduced set of decision variables, given in (P2):

$$\max_{\substack{\beta_m, \beta_s, \\ \mathbf{X}_m, \mathbf{X}_k}} u_m^{(2)}(\beta_m, \mathbf{X}_m) + \sum_{k=1}^{n} u_k^{(2)}(\beta_s, \mathbf{X}_k)$$

s.t.
$$\begin{cases} W_v \beta_m f_m^* r_{i,m} \geq \lambda_d L_d, \ i \in \mathcal{N}_u & \text{(3.14a)} \\[2mm] W_v \beta_m f_m^* r_{i,m} \geq c^{(min)}, \ i \in \mathcal{N}_a & \text{(3.14b)} \\[2mm] x_{i,m} \left[W_v \beta_m f_m^* r_{i,m} - c^{(min)} \right] \geq 0, \ i \in \mathcal{N}_k & \text{(3.14c)} \\[2mm] x_{i,m} \left[W_v \beta_m f_m^* r_{i,m} - \lambda_d L_d \right] \geq 0, \ i \in \mathcal{M}_k & \text{(3.14d)} \\[2mm] x_{i,k} \left[W_v \beta_s f_k^* r_{i,k} - c^{(min)} \right] \geq 0, \ i \in \mathcal{N}_k & \text{(3.14e)} \\[2mm] x_{i,k} \left[W_v \beta_s f_k^* r_{i,k} - \lambda_d L_d \right] \geq 0, \ i \in \mathcal{M}_k & \text{(3.14f)} \\[2mm] \sum_{j \in \{m,k\}} x_{i,j} = 1, \ i \in \mathcal{N}_k \cup \mathcal{M}_k & \text{(3.14g)} \\[2mm] x_{i,j} \in \{0, 1\}, \ i \in \mathcal{N}_k \cup \mathcal{M}_k, j \in \{m, k\} & \text{(3.14h)} \\[2mm] \beta_m + \beta_s = 1 & \text{(3.14i)} \\[2mm] \beta_m, \beta_s \in [0, 1] & \text{(3.14j)} \end{cases}$$

where $\mathbf{X}_m = \{x_{i,m} | i \in \mathcal{N}_k \cup \mathcal{M}_k, k \in \{1, 2, \ldots, n\}\}$, $\mathbf{X}_k = \{x_{i,k} | i \in \mathcal{N}_k \cup \mathcal{M}_k\}$, $u_m^{(2)}(\beta_m, \mathbf{X}_m)$ and $u_k^{(2)}(\beta_s, \mathbf{X}_k)$ are given by

$$u_m^{(2)}(\beta_m, \mathbf{X}_m) = \sum_{i \in \mathcal{N}_u \cup \mathcal{N}_a} \log\left(W_v \beta_m f_m^* r_{i,m}\right) + \sum_{k=1}^{n} \sum_{i \in \mathcal{N}_k \cup \mathcal{M}_k} x_{i,m} \log\left(W_v \beta_m f_m^* r_{i,m}\right)$$

$$\text{(3.15)}$$

and

$$u_k^{(2)}(\beta_s, \mathbf{X}_k) = \sum_{i \in \mathcal{N}_k \cup \mathcal{M}_k} x_{i,k} \log\left(W_v \beta_s f_k^* r_{i,k}\right). \qquad \text{(3.16)}$$

As the two sets of decision variables $\{\beta_m, \mathbf{X}_m\}$ and $\{\beta_s, \mathbf{X}_k\}$ are coupled through constraints (3.14g) and (3.14i), (P2) cannot be decoupled in the same way as (P1$'$). The simplified problem (P2) is a mixed-integer combinatorial problem with the binary variable set $\{x_{i,j}\}$, which is difficult to solve. Therefore, in the next section, we further transform (P2) to a tractable form for optimal solutions.

3.2.4 Problem Transformation for Partial Optimal Solutions

To solve (P2) in a tractable way, binary variables $\{x_{i,j}\}$ in (P2) are first relaxed to real-valued variables $\{\widetilde{x_{i,j}}\}$ within the range $[0, 1]$. Variables $\{\widetilde{x_{i,j}}\}$ represent the fraction of time that device i is associated with B_m or B_k during each resource slicing period [5]. With variable relaxation, the objective function of (P2) becomes a summation of $u_m^{(2)}(\beta_m, \widetilde{\mathbf{X}}_m)$ and $\sum_{k=1}^{n} u_k^{(2)}\left(\beta_s, \widetilde{\mathbf{X}}_k\right)$, where $\widetilde{\mathbf{X}}_m = \{\widetilde{x_{i,m}} | i \in \mathcal{N}_k \cup \mathcal{M}_k, k \in \{1, 2, \ldots, n\}\}$ and $\widetilde{\mathbf{X}}_k = \{\widetilde{x_{i,k}} | i \in \mathcal{N}_k \cup \mathcal{M}_k\}$. In Proposition 4, we provide the bi-concavity property of $u_m^{(2)}(\beta_m, \widetilde{\mathbf{X}}_m)$ and $\sum_{k=1}^{n} u_k^{(2)}\left(\beta_s, \widetilde{\mathbf{X}}_k\right)$, based on Definitions 1 and 2.

Definition 1 Suppose set Y is expressed as the Cartesian product of two subsets $A \in \mathbf{R}^{\mathbf{m}}$ and $B \in \mathbf{R}^{\mathbf{n}}$, i.e., $Y = A \times B$. Then, Y is called a biconvex set on $A \times B$, if A is a convex subset for any given $b \in B$, and B is also a convex subset for any given $a \in A$.

Definition 2 Function $\mathcal{F} : Y \rightarrow \mathbf{R}$ is defined on a biconvex set $Y = A \times B$, where $A \in \mathbf{R}^{\mathbf{m}}$ and $B \in \mathbf{R}^{\mathbf{n}}$. Then, $\mathcal{F}(A, B)$ is called a biconcave (biconvex) function if it is a concave (convex) function on subset A for any given $b \in B$, and it is also a concave (convex) function on subset B for any given $a \in A$.

Proposition 4 *Terms* $u_m^{(2)}(\beta_m, \widetilde{\mathbf{X}}_m)$ *and* $\sum_{k=1}^{n} u_k^{(2)}\left(\beta_s, \widetilde{\mathbf{X}}_k\right)$ *are (strictly) biconcave functions on the biconvex decision variable set* $\{\beta_m, \beta_s\} \times \{\widetilde{\mathbf{X}}_m, \widetilde{\mathbf{X}}_k, k \in \{1, 2, \ldots, n\}\}$.

The proof of Proposition 4 is given in Appendix D. Problem (P2) is then transformed to (P3):

$$\max_{\substack{\beta_m, \beta_s, \\ \widetilde{\mathbf{X}}_m, \widetilde{\mathbf{X}}_k}} u_m^{(2)}(\beta_m, \widetilde{\mathbf{X}}_m) + \sum_{k=1}^{n} u_k^{(2)}(\beta_s, \widetilde{\mathbf{X}}_k)$$

$$\text{s.t.}\begin{cases} W_v \beta_m \widetilde{f_m^*} r_{i,m} - \lambda_d L_d \geq 0, & i \in \mathcal{N}_u & (3.17a) \\[2mm] W_v \beta_m \widetilde{f_m^*} r_{i,m} - c^{(min)} \geq 0, & i \in \mathcal{N}_a & (3.17b) \\[2mm] \widetilde{x_{i,m}} \left[W_v \beta_m \widetilde{f_m^*} r_{i,m} - c^{(min)} \right] \geq 0, & i \in \mathcal{N}_k & (3.17c) \\[2mm] \widetilde{x_{i,m}} \left[W_v \beta_m \widetilde{f_m^*} r_{i,m} - \lambda_d L_d \right] \geq 0, & i \in \mathcal{M}_k & (3.17d) \\[2mm] \widetilde{x_{i,k}} \left[W_v \beta_s \widetilde{f_k^*} r_{i,k} - c^{(min)} \right] \geq 0, & i \in \mathcal{N}_k & (3.17e) \\[2mm] \widetilde{x_{i,k}} \left[W_v \beta_s \widetilde{f_k^*} r_{i,k} - \lambda_d L_d \right] \geq 0, & i \in \mathcal{M}_k & (3.17f) \\[2mm] \sum_{j \in \{m,k\}} \widetilde{x_{i,j}} = 1, & i \in \mathcal{N}_k \cup \mathcal{M}_k & (3.17g) \\[2mm] \widetilde{x_{i,j}} \in [0,1], & i \in \mathcal{N}_k \cup \mathcal{M}_k, j \in \{m,k\} & (3.17h) \\[2mm] \beta_m + \beta_s = 1 & & (3.17i) \\[2mm] \beta_m, \beta_s \in [0,1]. & & (3.17j) \end{cases}$$

The objective function in (P3) is a nonnegative summation of two (strictly) biconcave functions, which is also (strictly) biconcave [18]. Note that $\widetilde{f_m^*}$ and $\widetilde{f_k^*}$ are f_m^* and f_k^* with $x_{i,m}$ and $x_{i,k}$ substituted by $\widetilde{x_{i,m}}$ and $\widetilde{x_{i,k}}$. Moreover, if all the constraint functions in (P3) are written in a standard form, constraints (3.17a) and (3.17b) represent linear inequality constraint functions, and constraints (3.17g) and (3.17i) represent affine equality constraint functions, with respect to the set of decision variables. However, constraints (3.17c)–(3.17f) are non-convex constraint functions. Constraints (3.17c) and (3.17d) actually indicate that if any $i \in \mathcal{N}_k \cup \mathcal{M}_k$ is associated with B_m, the following inequalities need to be satisfied,

$$\sum_{k=1}^{n} \sum_{l \in \mathcal{N}_k \cup \mathcal{M}_k} \widetilde{x_{l,m}} \leq \frac{W_v \beta_m r_{i,m}}{c^{(min)}} - N_u - N_a, i \in \mathcal{N}_k \quad (3.18)$$

and

$$\sum_{k=1}^{n} \sum_{l \in \mathcal{N}_k \cup \mathcal{M}_k} \widetilde{x_{l,m}} \leq \frac{W_v \beta_m r_{i,m}}{\lambda_d L_d} - N_u - N_a, i \in \mathcal{M}_k. \quad (3.19)$$

If device $i(\in \mathcal{N}_k \cup \mathcal{M}_k)$ is associated with B_k, constraints (3.17e) and (3.17f) are equivalent to

$$\sum_{l \in \mathcal{N}_k \cup \mathcal{M}_k} \widetilde{x_{l,k}} \leq \frac{W_v \beta_s r_{i,k}}{c^{(min)}}, \quad i \in \mathcal{N}_k \quad (3.20)$$

and

$$\sum_{l \in \mathcal{N}_k \cup \mathcal{M}_k} \widetilde{x_{l,k}} \leq \frac{W_v \beta_s r_{i,k}}{\lambda_d L_d}, \quad i \in \mathcal{M}_k. \tag{3.21}$$

Therefore, to make (P3) tractable, we simplify (P3) to (P3$'$), by substituting (3.17c)–(3.17f) with (3.18)–(3.21), respectively:

$$(\text{P3}') : \max_{\substack{\beta_m, \beta_s, \\ \widetilde{\mathbf{X}}_m, \widetilde{\mathbf{X}}_k}} u_m^{(2)}(\beta_m, \widetilde{\mathbf{X}}_m) + \sum_{k=1}^{n} u_k^{(2)}(\beta_s, \widetilde{\mathbf{X}}_k)$$

s.t. (3.17a), (3.17b), (3.17g)–(3.17j), (3.18)–(3.21).

Compared with (3.17a)–(3.17d) in (P3), constraints (3.17a), (3.17b), (3.18) and (3.19) in (P3$'$) provide the lowest upper bound on the number of category II devices associated with B_m. This lowest upper bound is tight since the communication distance between the MBS and any device located in the coverage of an SBS is much longer than the location differences among MTDs and MUs in the same SBS. Therefore, the differences of $r_{i,m}$ among the end devices are small. Similarly, compared with constraints (3.17e) and (3.17f) in (P3), constraints (3.20) and (3.21) in (P3$'$) provide the lowest upper bound on the number of category II devices associated with $B_k, k \in \{1, 2, \ldots, n\}$. Without changing the optimal solutions for (P3), the simplified constraints (3.18)–(3.21) in (P3$'$) indicate a set of conservative limits on maximum numbers of category II MTDs and MUs that can be associated with B_m and $B_k, k \in \{1, 2, \ldots, n\}$.

Note that (P3$'$) is a standard biconcave maximization problem due to the biconcave objective function and biconvex constraint functions with respect to the biconvex decision variable set $\{\beta_m, \beta_s\} \times \{\widetilde{\mathbf{X}}_m, \widetilde{\mathbf{X}}_k\}$ [18]. To solve (P3$'$), an *alternative concave search* (ACS) algorithm is designed by leveraging the biconcavity of the problem. The detailed ACS algorithm is presented in Algorithm 2. As stated in Proposition 5, due to the properties of (P3$'$), Algorithm 2 converges to a set of *partial optimal solutions*. The definition of a partial optimal solution for (P3$'$) is given in Corollary 1, based on Proposition 5 and Theorem 4.7 in [18].

Proposition 5 *Algorithm 2 converges, because (1) both $\{\beta_m, \beta_s\}$ and $\{\widetilde{\mathbf{X}}_m, \widetilde{\mathbf{X}}_k\}$ are closed sets, and the objective function of (P3$'$) is continuous on its domain; (2) Given the set of accumulation points,[1] $\{\beta_m^{(t)}, \beta_s^{(t)}, \widetilde{\mathbf{X}}_m^{(t)}, \widetilde{\mathbf{X}}_k^{(t)}\}$, at the beginning of tth iteration, the optimal solutions at the end of tth iteration (at the beginning of $(t + 1)$th iteration), i.e., $\{\beta_m^{(t+1)}, \beta_s^{(t+1)}, \widetilde{\mathbf{X}}_m^{(t+1)}, \widetilde{\mathbf{X}}_k^{(t+1)}\}$, are unique solutions.*

The proof of Proposition 5 is given in Appendix E.

---

[1] An accumulation point set for the ACS algorithm denotes the set of optimal solutions at the beginning of tth (for any $t > 0$) iteration.

Corollary 1 *Algorithm 2 converges to a set of optimal solutions, called partial optimums* $\{\beta_m^*, \beta_s^*, \widetilde{\mathbf{X}}_m^*, \widetilde{\mathbf{X}}_k^*\}$ *where* $\widetilde{\mathbf{X}}_m^* = \{\widetilde{x_{i,m}}^* | i \in \mathcal{N}_k \cup \mathcal{M}_k, k \in \{1, 2, \ldots, n\}\}$ *and* $\widetilde{\mathbf{X}}_k^* = \{\widetilde{x_{i,k}}^* | i \in \mathcal{N}_k \cup \mathcal{M}_k\}$, *which satisfy*

$$
u_m^{(2)}(\beta_m^*, \widetilde{\mathbf{X}}_m^*) + \sum_{k=1}^{n} u_k^{(2)}(\beta_s^*, \widetilde{\mathbf{X}}_k^*)
$$
$$
\geq u_m^{(2)}(\beta_m, \widetilde{\mathbf{X}}_m^*) + \sum_{k=1}^{n} u_k^{(2)}(\beta_s, \widetilde{\mathbf{X}}_k^*), \quad \forall \beta_m, \beta_s \in [0, 1] \tag{3.22}
$$

and

$$
u_m^{(2)}(\beta_m^*, \widetilde{\mathbf{X}}_m^*) + \sum_{k=1}^{n} u_k^{(2)}(\beta_s^*, \widetilde{\mathbf{X}}_k^*)
$$
$$
\geq u_m^{(2)}(\beta_m^*, \widetilde{\mathbf{X}}_m) + \sum_{k=1}^{n} u_k^{(2)}(\beta_s^*, \widetilde{\mathbf{X}}_k), \quad \forall \widetilde{x_{i,m}}, \widetilde{x_{i,k}} \in [0, 1]. \tag{3.23}
$$

In Algorithm 2, the main logical flow is to iteratively solve for optimal radio resource slicing ratios and optimal BS-device association patterns, $\{\beta_m^*, \beta_s^*, \widetilde{\mathbf{X}}_m^*, \widetilde{\mathbf{X}}_k^*\}$. In each iteration, given a set of optimal values of β_m and β_s from the previous iteration, (P3$'$) is solved for a set of optimal BS-device association patterns $\{\widetilde{\mathbf{X}}_m^\dagger, \widetilde{\mathbf{X}}_k^\dagger\}$ and then, given $\{\widetilde{\mathbf{X}}_m^\dagger, \widetilde{\mathbf{X}}_k^\dagger\}$, (P3$'$) is solved again for updated optimal resource slicing ratios $\{\beta_m^\dagger, \beta_s^\dagger\}$. At this point, the current iteration ends, and the stopping criterion for Algorithm 2 is checked, i.e., whether the difference between the objective function values at the end of current iteration and at the end of previous iteration is smaller than a predefined threshold (set as a small value). If the stopping criterion is met, the set of optimal solutions for current iteration converge to the final optimal solution set $\{\beta_m^*, \beta_s^*, \widetilde{\mathbf{X}}_m^*, \widetilde{\mathbf{X}}_k^*\}$. Otherwise, the next round of iteration begins until the algorithm converges. For each pair of optimal solutions $\{\widetilde{x_{i,m}}^*, \widetilde{x_{i,k}}^*\}$, we let the larger one equal 1 and the smaller one equal 0 to obtain the optimal solutions $\{x_{i,m}^*, x_{i,k}^*\}$ and ensure every end device is associated with one BS during each resource slicing period. Simulation results in Sect. 3.2.5.2 show accuracy of the variable relaxation in solving (P2).

The computational complexity of Algorithm 2 is calculated as follows: In t th iteration, given the set of resource slicing ratios, $\{\beta_m^{(t)}, \beta_s^{(t)}\}$, the convex optimization problem (P3$'$) is solved for the BS-device association patterns, $\{\widetilde{\mathbf{X}}_m^{(t+1)}, \widetilde{\mathbf{X}}_k^{(t+1)}\}$, where the number of decision variables is $2\sum_{k=1}^{n}(N_k + M_k)$; Then, given $\{\widetilde{\mathbf{X}}_m^{(t+1)}, \widetilde{\mathbf{X}}_k^{(t+1)}\}$, (P3$'$) with 2 decision variables is solved again for the resource slicing ratios, $\{\beta_m^{(t+1)}, \beta_s^{(t+1)}\}$ at the end of t th iteration. Therefore, in each iteration, both convex optimization problems are solved sequentially by

---

Algorithm 2: The ACS algorithm for solving (P3$'$)

Input : Input parameters for (P3$'$), stopping criterion δ, iteration limit $N_{\rm m}$, a candidate set
        C of initial values for $\{\beta_m, \beta_s\}$.

Output: Optimal slicing ratios, $\{\beta_m^*, \beta_s^*\}$; Optimal association patterns, $\{\widetilde{\mathbf{X}}_m{}^*, \widetilde{\mathbf{X}}_k{}^*\}$.

1 **Step 1**: Select a pair of initial values for $\{\beta_m, \beta_s\}$ from C, denoted by $\{\beta_m^{(t)}, \beta_s^{(t)}\}$ where
      $t = 0$; Let $\mathcal{U}^{(t)}$ denote the maximum objective function value, with optimal decision
      variables $\{\beta_m^{(t)}, \beta_s^{(t)}, \widetilde{\mathbf{X}}_m^{(t)}, \widetilde{\mathbf{X}}_k^{(t)}\}$, at the beginning of tth iteration;

2 **Step 2**: $\mathcal{U}^{(0)} \leftarrow 0$;

3 **do**

4      $\{\widetilde{\mathbf{X}}_m{}^{\dagger}, \widetilde{\mathbf{X}}_k{}^{\dagger}\} \leftarrow$ solving (P3$'$) given $\{\beta_m^{(t)}, \beta_s^{(t)}\}$;

5   **if** *No feasible solutions for* (P3$'$) **then**

6   Go to **Step 1** until no feasible solutions found with initial values in C;

7   Stop and no optimal solutions under current network conditions;

8   **else**

9          $\{\widetilde{\mathbf{X}}_m^{(t+1)}, \widetilde{\mathbf{X}}_k^{(t+1)}\} \leftarrow \{\widetilde{\mathbf{X}}_m{}^{\dagger}, \widetilde{\mathbf{X}}_k{}^{\dagger}\}$;

10         $\{\beta_m^{\dagger}, \beta_s^{\dagger}\} \leftarrow$ solving (P3$'$) given $\{\widetilde{\mathbf{X}}_m^{(t+1)}, \widetilde{\mathbf{X}}_k^{(t+1)}\}$;

11   **if** *No feasible solutions for* (P3$'$) **then**

12   Go to **Step 1** until no feasible solutions found with initial values in C;

13   Stop and no optimal solutions under current network conditions;

14   **else**

15           $\{\beta_m^{(t+1)}, \beta_s^{(t+1)}\} \leftarrow \{\beta_m^{\dagger}, \beta_s^{\dagger}\}$;

16   Obtain maximum objective function value $\mathcal{U}^{(t+1)}$ at the end of tth iteration,
           with $\{\beta_m^{(t+1)}, \beta_s^{(t+1)}, \widetilde{\mathbf{X}}_m^{(t+1)}, \widetilde{\mathbf{X}}_k^{(t+1)}\}$.

17   **end**

18        $t \leftarrow t + 1$;

19   **end**

20 **while** $\|\mathcal{U}^{(t)} - \mathcal{U}^{(t-1)}\| \geq \delta$ **or** $N_{\rm m}$ is not reached;

---

using interior-point methods [19, 20], and thus the time complexity upper bound of
Algorithm 2 is $O\left[N_{\rm m}\left(\sum\limits_{k=1}^{n} (N_k + M_k) \right)^4 \right]$, where n is the number of small cells
within the macro-cell.

3.2.5 Simulation Results

In this subsection, simulation results are presented to demonstrate the effectiveness
of our proposed dynamic radio resource slicing framework. All the simulations
are carried out using MATLAB. In a two-tier wireless network with 1 macro-cell
underlaid by 4 small cells, deployment locations of the MBS and SBSs are fixed,
and the distance between the MBS and each of the SBSs are set as 400 m. The
downlink transmit power on the MBS is set to 40 dBm with the communication
coverage radius of 600 m, whereas each SBS has identical transmit power of 30 dBm

Table 3.1 System parameters

Parameters	Values
Aggregate bandwidth resources (W_v)	20 MHz [5]
Background noise power (σ^2)	−104 dBm
Data packet size (L_d)	9000 bits
Machine-type packet size (L_a)	2000 bits [22]
Machine-type packet delay bound (D_{max})	100 ms [22]
Delay bound violation probability (ε)	10^{-3} [22]
Stopping criterion (δ)	0.01
Iteration limit (N_m)	1000 rounds

with the coverage radius of 200 m [21]. All devices are randomly scattered in the coverages of the MBS and SBSs, and each small cell is set an equal number of category II MTDs and category II MUs, denoted by N_s and M_s, respectively. In each small cell, M_s is set to 10. For wireless propagation models, we use $L_m(z) = -30 - 35 \log_{10}(z)$ and $L_s(z) = -40 - 35 \log_{10}(z)$ to describe the downlink path loss channel gains for the macro-cell and each small cell, respectively, where z is the distance between a BS and a device. λ_d and λ_a are set as 20 packet/s and 5 packet/s, respectively. Each simulation point is obtained by averaging over 50 location distribution samples of MUs and MTDs. Other important system parameters for simulations are summarized in Table 3.1.

We first demonstrate through extensive simulations the robustness of the proposed radio resource slicing framework, where the optimal slicing ratios are obtained and dynamically adjusted with low complexity. Then, the proposed slicing framework is compared with an SINR-maximization (SINR-max) based network-level resource slicing scheme [5], where radio resources are shared among BSs and each device is associated with the BS providing highest downlink SINR, and a service-level resource slicing scheme [5], i.e., radio resources are pre-configured on each BS, and are sliced for different service groups under the BS.

3.2.5.1 Optimal Radio Resource Slicing Ratios

The solutions for radio resource slicing ratio, β_s, at each iteration of Algorithm 2 are shown in Fig. 3.5a and b. In Fig. 3.5a, given the numbers of MUs and MTDs in the macro cell and each small cell, β_s converges to the same optimal solution, regardless of the location distribution for MUs and MTDs. The same pattern is observed under a different network load condition in Fig. 3.5b. This observation demonstrates the robustness of the proposed resource slicing framework, where the optimal slicing ratios remain unchanged with end device location variations in each cell. Therefore, the slicing ratios are adjusted in a large timescale, which significantly reduces the amount of communication overhead for updating the slicing ratios.

Fig. 3.5 Radio resource slicing ratio (β_s) for an SBS during each iteration of Algorithm 2. (**a**) $N_u = N_a = 25$, $N_s = 40$, $M_s = 10$. (**b**) $N_u = N_a = 100$, $N_s = 90$, $M_s = 10$

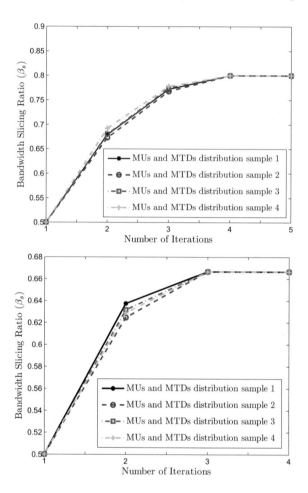

3.2.5.2 Performance Comparison

In Fig. 3.6, optimal resource slicing ratios are compared between the proposed slicing framework and the SINR-max based network-level slicing scheme for different values of N_u, N_a, and N_s. In the SINR-max based slicing scheme, since each device is always connected to the BS providing the highest SINR, the BS-device association patterns frequently change upon variations of end device locations, and radio resources are adjusted accordingly to adapt to the varying traffic load under each BS. Figure 3.6 shows that the optimal slicing ratio in the SINR-max based slicing scheme fluctuates with MU and MTD distributions. In comparison, the proposed slicing framework is more robust with network dynamics, and the slicing ratios are updated in a much larger timescale with reduced communication overhead.

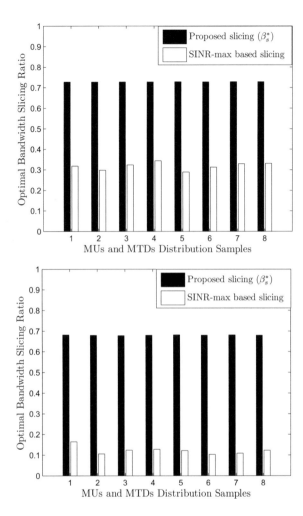

Fig. 3.6 Comparison of bandwidth slicing ratios between the proposed slicing framework and the SINR-max based slicing scheme. (**a**) $N_u = N_a = 75$, $N_s = 90$, $M_s = 10$. (**b**) $N_u = N_a = 100$, $N_s = 140$, $M_s = 10$

Next, we compare the performance of the proposed resource slicing framework with the service-level resource slicing scheme in Fig. 3.7, where it can be seen that for the service-level resource slicing scheme, with the increase of N_s, more and more MTDs and MUs are connected to the MBS to improve the overall resource utilization. As a result, MTDs and MUs need to frequently change their associations with BSs, which inevitably increases the communication overhead. In contrast, for the proposed resource slicing framework, the radio resources are dynamically adjusted on BSs in response to the network traffic variations, leading to more flexible and cost-effective resource orchestration with reduced the communication cost.

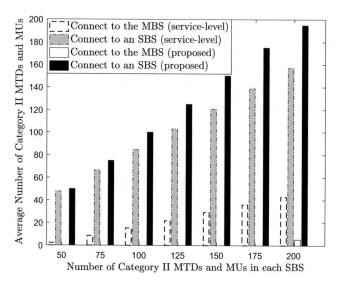

Fig. 3.7 Average number of category II MTDs and MUs connecting to either the MBS or their home SBSs ($N_u = N_a = 50$)

The aggregate network utilities achieved by different resource slicing schemes, are also compared with variations of N_u, N_a and N_s in Fig. 3.8a, b. For the proposed slicing framework, we also evaluate the effect of the approximations (i.e., the variable relaxation) made for solving (P2). It is clear that the network utility achieved by the set of fractional BS-device optimal association patterns $\{\widetilde{x_{i,m}}^*, \widetilde{x_{i,k}}^*\}$ matches closely with the one achieved by the exact binary optimal solutions $\{x_{i,m}^*, x_{i,k}^*\}$ after the rounding operation. Moreover, through radio resource slicing among BSs, the overall network resource utilization are significantly improved. Thus, it is observed that our proposed resource slicing framework achieves higher network utility than both the SINR-max-based network-level resource slicing scheme and the service-level resource slicing scheme. Figure 3.9 demonstrates the adaptiveness of our proposed slicing framework where the optimal radio resource slicing ratio β_s^* is dynamically updated with the number of devices in each SBS, compared with a fixed resource configuration for the service-level slicing scheme.

3.3 Summary

In this chapter, we have presented an E2E resource slicing framework for both 5G core and wireless networks. For a 5G core network, the DR-GPS is employed as the bi-resource slicing scheme to achieve dominant-resource fairness and high resource utilization with minimized queueing delay for packet transmission; For a 5G wireless access network, a radio resource slicing problem is studied to optimize

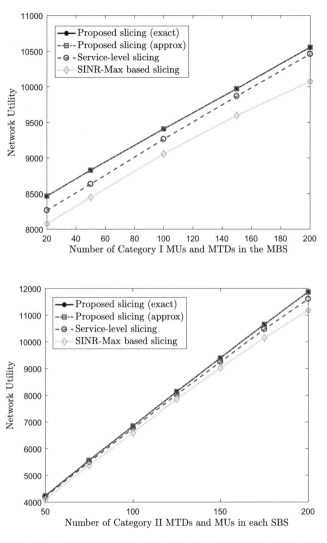

Fig. 3.8 Comparison of aggregate network utility (**a**) with respect to the number of category I MUs and MTDs ($N_u = N_a$, $N_s = 140$, $M_s = 10$) and (**b**) with respect to the number of category II MTDs and MUs ($N_u = N_a = 50$)

a set of resource slicing ratios and BS-device association patterns for maximizing the overall network utility with differentiated QoS guarantee for both data and M2M services. The amount of radio resources on each BS are also dynamically adjusted by adapting to the variation of network traffic load. Extensive simulation results demonstrate the effectiveness and advantages of the proposed E2E network slicing framework for 5G networks compared with benchmark slicing schemes.

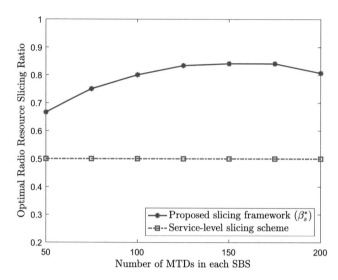

Fig. 3.9 Adaptation of the proposed slicing framework with network load variations ($N_u = N_a = 50$)

References

1. W. Wang, B. Liang, B. Li, Multi-resource generalized processor sharing for packet processing, in *Proc. ACM IWQoS*, 2013, pp. 1–10
2. A. Ghodsi, M. Zaharia, B. Hindman, A. Konwinski, S. Shenker, I. Stoica, Dominant resource fairness: fair allocation of multiple resource types, in *Proc. USENIX NSDI*, 2011, pp. 323–336
3. A. Ksentini, N. Nikaein, Toward enforcing network slicing on RAN: flexibility and resources abstraction. IEEE Commun. Mag. **55**(6), 102–108 (2017)
4. O. Sallent, J. Perez-Romero, R. Ferrus, R. Agusti, On radio access network slicing from a radio resource management perspective. IEEE Wirel. Commun. **24**(5), 166–174 (2017)
5. C. Liang, F.R. Yu, H. Yao, Z. Han, Virtual resource allocation in information-centric wireless networks with virtualization. IEEE Trans. Veh. Technol. **65**(12), 9902–9914 (2016)
6. A.A. Gebremariam, M. Chowdhury, A. Goldsmith, F. Granelli, Resource pooling via dynamic spectrum-level slicing across heterogeneous networks, in *Proc. IEEE CCNC' 17*, 2017, pp. 818–823
7. X. Foukas, M.K. Marina, K. Kontovasilis, Orion: RAN slicing for a flexible and cost-effective multi-service mobile network architecture, in *Proc. ACM MobiCom '17*, 2017, pp. 127–140
8. D. Wu, R. Negi, Effective capacity-based quality of service measures for wireless networks. ACM Mobile Netw. Appl. **11**(1), 91–99 (2006)
9. T.P.C. de Andrade, C.A. Astudillo, N.L.S. da Fonseca, Allocation of control resources for machine-to-machine and human-to-human communications over LTE/LTE-A networks. IEEE Internet Things J. **3**(3), 366–377 (2016)
10. E. Soltanmohammadi, K. Ghavami, M. Naraghi-Pour, A survey of traffic issues in machine-to-machine communications over LTE. IEEE Internet Things J. **3**(6), 865–884 (2016)
11. Y. Liu, C. Yuen, X. Cao, N.U. Hassan, J. Chen, Design of a scalable hybrid MAC protocol for heterogeneous M2M networks. IEEE Internet Things J. **1**(1), 99–111 (2014)
12. L. Li, M. Pal, Y.R. Yang, Proportional fairness in multi-rate wireless LANs, in *Proc. IEEE INFOCOM' 08* (2008)

13. P. Caballero et al., Multi-tenant radio access network slicing: statistical multiplexing of spatial loads. IEEE/ACM Trans. Netw. **25**(5), 3044–3058 (2017)
14. A. Aijaz, M. Tshangini, M.R. Nakhai, X. Chu, A.H. Aghvami, Energy-efficient uplink resource allocation in LTE networks with M2M/H2H co-existence under statistical QoS guarantees. IEEE Trans. Commun. **62**(7), 2353–2365 (2014)
15. C.Y. Oh, D. Hwang, T.J. Lee, Joint access control and resource allocation for concurrent and massive access of M2M devices. IEEE Trans. Wireless Commun. **14**(8), 4182–4192 (2015)
16. D. Wu, R. Negi, Effective capacity: a wireless link model for support of quality of service. IEEE Trans. Wirel. Commun. **2**(4), 630–643 (2003)
17. Q. Ye, W. Zhuang, S. Zhang, A. Jin, X. Shen, X. Li, Dynamic radio resource slicing for a two-tier heterogeneous wireless network. IEEE Trans. Veh. Technol. **67**(10), 9896–9910 (2018)
18. J. Gorski, F. Pfeuffer, K. Klamroth, Biconvex sets and optimization with biconvex functions: a survey and extensions. Math. Method. Oper. Res. **66**(3), 373–407 (2007)
19. S.P. Boyd, L. Vandenberghe, *Convex Optimization* (Cambridge University Press, Cambridge, 2004)
20. A. Ben-Tal, A. Nemirovski, *Lectures on Modern Convex Optimization: Analysis, Algorithms, and Engineering Applications* (Society for Industrial and Applied Mathematics (SIAM), 2001)
21. S. Zhang, J. Gong, S. Zhou, Z. Niu, How many small cells can be turned off via vertical offloading under a separation architecture?. IEEE Trans. Wirel. Commun. **14**(10), 5440–5453 (2015)
22. 3rd Generation Partnership Project; Technical Specification Group Services and System Aspects; Feasibility Study on New Services and Markets Technology Enablers for Critical Communications; Stage 1 (Release 14) , in *3GPP TR 22.862 V14.1.0*, 2016, pp. 1–31

Chapter 4
Transport-Layer Protocol Customization for 5G Core Networks

4.1 SDN/NFV-Based Transmission Protocol Customization

With rapid development of networking technologies, the 5G communication net-works are foreseen to accommodate diversified data-hungry mobile services and delay/reliability sensitive IoT applications. However, due to current distributed and ossified core network architecture with limited computing and transmission resources on network elements (e.g., network servers, and forwarding devices), it is challenging to guarantee differentiated QoS for diverse services [1, 2]. Therefore, developing an efficient transport-layer protocol is imperative to mitigate network congestion and enhance E2E QoS satisfaction in a fine-grained way at the network operation stage. Transmission control protocol (TCP) [3, 4] is a typical transport-layer protocol that is widely used for reliable E2E transmissions in packet switched networks. In TCP, through a three-way handshake mechanism during the connection establishment phase, a pair of end hosts establish a two-way communication connec-tion for data transmission. The sending host also uses retransmission timeout (RTO) and fast retransmission with congestion window adjustment for lost packet recovery and congestion control. To support increased data traffic volume and different levels of E2E delay requirements from diversified services, more and more network elements (i.e., servers and switches) are placed into the network which expands the network deployment and operational cost in a large scale. Due to imbalanced traffic load, resources at some network locations are underutilized, whereas other network segments probably undergo high traffic congestion. To improve E2E QoS performance for diversified services, existing studies propose enhanced transport-layer protocols, including false fast retransmission avoidance [5, 6], accurate RTO estimation [7, 8], and efficient congestion control [9, 10]. However, due to the distributed and ossified network architecture, current transport-layer protocols, e.g., TCP and user datagram protocol (UDP), only achieve best-effort E2E performance [11]. In TCP, a sending host records a round-trip time (RTT) for every transmitted data packet for packet loss detection and congestion control; A three duplicate

Q. Ye, W. Zhuang, *Intelligent Resource Management for Network Slicing in 5G and Beyond*, Wireless Networks, https://doi.org/10.1007/978-3-030-88666-0_4

acknowledgments (ACKs) mechanism is applied for fast retransmission of lost packets. However, the TCP congestion detection is performed only at a sending host based on ACK feedback from its receiving end. Without in-network congestion awareness, a long response delay can be experienced for packet retransmission and congestion control.

With the help of SDN, the control intelligence is decoupled from the data plane to centrally manage data traffic over a core network for traffic load balancing and network congestion alleviation [12–14]. In addition, traffic flows of certain type of service from different end hosts are aggregated at the ingress edge node of the core network and traverse a sequence of network functions, e.g., firewall and IDS, for packet-level processing to fulfill the service requirements [15]. With NFV, network/service functions are softwarized as VNFs/virtual service functions (VSFs) which are instantiated in generalized commodity servers [16, 17]. The NFV enables flexible function instantiation on different NFV nodes with reduced CapEx and OpEx.

The SDN/NFV networking architecture provides an open programmable physical substrate network for instantiating different VNFs on NFV nodes and customizing the traffic routing paths traversing the VNFs with properly allocated processing and transmission resources to achieve high E2E performance [18, 19]. However, the VNF placement and routing path configuration with associated resource allocation are performed in a large timescale (e.g. minutes or hours), which do not capture the small-timescale traffic burstiness. To better accommodate traffic variations from different services, more fine-grained transmission control is required to reduce the level of traffic congestion. Some in-network protocol functionalities can be enabled to realize early congestion detection and reaction with reduced delay and overhead, such as in-path packet caching and retransmission functions. Compared with a conventional TCP network, these in-network functionalities require computing, caching, and higher protocol-layer packet analyzing capabilities. However, how to properly balance congestion control with E2E QoS performance remains an important but challenging issue. Moreover, the in-network control needs to be customized for service-oriented QoS provisioning. In the following two sections, we present SDN/NFV-based adaptive transmission protocols to support both time-critical and throughput-oriented VoD streaming services.

4.2 Protocol Customization for Time-Critical Services

In this section, we develop an SDN-based adaptive transmission protocol (SDATP) supporting a time-critical service that requires high reliability and low latency, where in-path packet caching and retransmission functionalities are enabled for in-network congestion control. In fact, how the caching and retransmission functionalities are activated in the network significantly affects the E2E QoS performance. More enabled caching functionalities reduce the overflow probability of each individual caching buffer, but lead to less efficient caching resource usage. Given a number of activated caching nodes, the locations of these caching functionalities also affect QoS provisioning. For an E2E network path, if a caching functionality is activated near the receiving end, a high performance gain (e.g., reduced packet retransmission hops) is reached for each lost packet. However, the caching effectiveness is also compromised due to the lowered chances of having lost packets retransmitted by in-path caching nodes for early recovery. Thus, the performance should be optimized by balancing the trade-off between caching effectiveness and resource utilization.

Therefore, in this section, the caching node placement and packet caching probability are jointly considered to improve the caching efficiency and, at the same time, minimize the E2E packet delay for time-critical services (e.g., MTC services for industrial automation). This joint problem is formulated as a mixed-integer nonlinear program (MINLP), and the dependency between packet caching probabilities and locations of caching node placement poses technical challenges in solving the problem. For a tractable solution, we simplify the problem by reducing the number of decision variables, which is achieved by taking into consideration the feature of in-path caching. Then, a low complexity heuristic algorithm is developed to solve the simplified problem. In comparison with benchmark schemes, the proposed probabilistic caching policy achieves the minimum number of retransmission hops with fastest packet retransmissions.

4.2.1 Network Model

We consider multiple virtual networks accommodated on a 5G physical network substrate for supporting diverse services. Data traffic from a set of E2E connections belonging to one service type are aggregated at virtual switches[1] (i.e., vSwitches) at the edge of the core network as one traffic flow. Thus, the edge vSwitch is equipped with the traffic aggregation and higher-layer network functionalities (e.g., header processing). Each E2E network path consists of three network portions: (1) from the source node to ingress edge vSwitch, (2) between two edge vSwitches, and (3) from the egress edge vSwitch to the destination node. To guarantee backward compatibility of protocols, the transmissions between end hosts and edge vSwitch are supervised by the TCP protocol. A customized transmission protocol is developed to enhance the data transmission performance between a pair of ingress and egress edge vSwitches for supporting a time-critical service. The protocol is operated at a slice level. We define one network *slice* as a combination of customized protocol elements, allocated resources, and the embedded virtual network topology to support one aggregated traffic flow of certain service between a

---

[1] A virtual switch refers to a softwarized network switch with virtualized resources to program higher-layer network functions, apart from the functionalities of traffic forwarding or traffic aggregation.

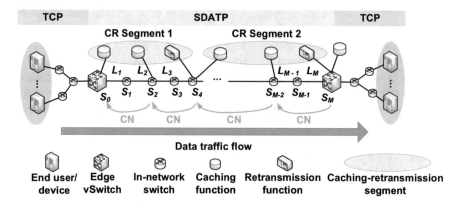

Fig. 4.1 An illustration of an embedded virtual network for time-critical service delivery

pair of edge vSwitches. Slices are differentiated and logically isolated to guarantee different levels of service requirements. Each slice has a unique *slice ID* which is a combination of source and destination edge switch IP addresses and a pair of port numbers for E2E service delivery.

An embedded virtual network with linear topology supporting a time-critical service is shown in Fig. 4.1, where a data traffic flow aggregated from a set of end users/devices belonging to the service traverses through the network. Let M denote the number of transmission links between edge vSwitches, and S_0 and S_M denote the ingress and egress edge vSwitches. The traversed path consists of $(M - 1)$ in-network vSwitches, denoted by $\{S_1, S_2, \ldots, S_{M-1}\}$, and M transmission links between S_0 and S_M, denoted by $\{L_1, L_2, \ldots, L_M\}$. Apart from all the vSwitches centrally managed by an SDN controller, there also exist conventional switches, with only packet forwarding capability, which is not shown in Fig. 4.1 for clarity. In the following, a switch refers to a vSwitch. Since services may require caching functions, both the two edge switches and the $(M - 1)$ in-network switches are equipped with caching buffers, the sizes of which are denoted as $\{B_0, B_1, \ldots, B_M\}$. For SDATP, caching functionalities are activated at $N (\leq M - 1)$ in-network switches, and the index for one caching node is the index of the corresponding network switch. The set of indexes for the caching nodes is denoted by $\mathbb{C}(N) = \{C_1, C_2, \ldots, C_N\}$. For example, we have $C_1 = 2$ and $C_2 = 4$ shown in Fig. 4.1. Since both edge switches (S_0 and S_M) are always activated with caching functions to cache all received packets, the caching policy is designed only for in-network switches S_m ($m = 1, 2, \ldots, M - 1$). Packet loss at the mth ($m = 1, 2, \ldots, M$) transmission hop includes packet loss due to transmission errors of link L_m and congestion at switch S_{m-1}. If congestion happens at S_M, it will be handled by TCP operating between S_M and the receiving node.

4.2.2 Network Functionalities

Consider that the supported service demands high reliability and low latency. Thus, lost packets need to be retransmitted. A set of on-demand in-network protocol functionalities can be activated for packet loss detection and packet retransmission, by triggering caching and retransmission functionalities at in-network switches.

4.2.2.1 Caching Function

A caching node indicates a network switch with activated caching function, which is equipped with pairs of data transmission and caching buffers for different aggregated traffic flows. When a node receives a data packet, it caches the packet with a certain probability, which is termed as *probabilistic caching*. The caching buffer is used to store cached packets, while the data transmission buffer is used to queue packets to be forwarded. Each network slice has a unique pair of data transmission and caching buffers.

Caching Release Function Due to the buffer capacity constraint, the caching buffer release function is necessary to avoid overflow, which is realized by transmitting a caching notification (CN) packet upstream from one caching node to its preceding one for releasing cached packets, as shown in Fig. 4.1. Time is partitioned into intervals with constant duration T (in the unit of second), and a caching node sends a CN packet to its preceding caching node in every T time interval.

Each CN packet contains the following information: (1) Packet type, indicating a caching notification packet; (2) Caching release window, containing the sequence numbers of successfully cached packets at a caching node during T. When a caching node receives a CN packet from its subsequent caching node, it releases the portion of cached packets according to the received caching release window; (3) Caching state information, the available buffer size of a caching node, indicating current caching node's packet caching capability. Such information is also used for in-network congestion control (to be discussed in Sect. 4.2.4.4).

4.2.2.2 Retransmission Function

A network switch with activated retransmission function is referred to as a retransmission node, with in-path packet loss detection and retransmission triggering functionalities. When a retransmission node detects packet loss, it triggers retransmission by sending a retransmission request to upstream caching nodes consecutively. If a requested data packet is discovered at a caching node, the packet will be retransmitted from that caching node. We refer to *one caching-retransmission (CR) segment* as a network segment, including the network switches/transmission links between two consecutive retransmission nodes, as shown in Fig. 4.1. The caching and retransmission functions within one CR segment cooperatively recover the packet loss occurred in the segment, and multiple segments can operate in parallel to reduce the retransmission delay.

In the virtual network establishment, the SDN controller makes decisions on where to activate caching or retransmission functionality. To minimize the maximum packet retransmission delay,[2] both the ingress and egress edge switches are equipped with the caching functionality, while the egress edge switch is also enabled retransmission functionality. Specifically, based on packet loss probability and available resources over each transmission hop, the policy of activating caching and retransmission functionalities is discussed in Sect. 4.2.5. If a packet is lost between two CR segments, the packet may be released by the caching nodes in preceding CR segment, while it has not been received by the caching nodes in the subsequent CR segment. Therefore, the packet loss between two CR segments requires retransmission from the source node, which leads to a longer delay than retransmission from in-path caching nodes. To ensure seamlessness in packet caching, we assume the retransmission node of one CR segment is also the caching node of its subsequent CR segment.

4.2.3 Traffic Model

Packet arrivals for the time-critical service with high-reliability and low-latency requirements at the ingress edge vSwitch are modeled as a Poisson process, with the consideration of temporal independence of access requests from different users [20, 21]. The packet arrival rate of the aggregated flow ranges from tens to a few hundreds packets per second [21]. In each time interval of T, the number of packets arriving at edge switch S_0, denoted by Y, follows a Poisson distribution, with mean number of packet arrivals denoted by λ over T. It is proved in [22] that bounds of the median of Y satisfy

$$\lambda - \ln(2) \le median\,(Y) < \lambda + \frac{1}{3}. \tag{4.1}$$

With λ much greater than 1, we have $median\,(Y) \approx \lambda$, i.e., $Pr\,\{Y \le \lambda\} \approx 0.5$.

4.2.4 Customized Protocol Design

A customized protocol for a time-critical service with high-reliability and low-latency requirements is proposed, where the protocol function elements include

---

[2] Packet retransmission delay is the time duration from the instant that a retransmission request packet is sent (after detecting packet loss) to the instant that a retransmitted packet is received by the requesting node.

connection establishment, data transmission, in-path caching-based packet retransmission, and congestion control. The objective is to minimize E2E packet delay by activating these elements on demand. We also develop an optimization framework to determine the probabilistic caching policy, including the optimal number, caching function placement, and packet caching probabilities.

4.2.4.1 Connection Establishment

For conventional TCP used in a packet switched network, a *three-way* handshake mechanism is required for connection establishment to ensure the reachability of an E2E path before data transmission. However, based on the SDN architecture, the three-way handshake process can be simplified as follows:

1. With global network information, an SDN controller checks the E2E path availability, which is more efficient than distributed signaling exchange;
2. The SDN controller assigns the initial sequence number and acknowledgment number to end hosts to prepare for data transmissions;
3. During the connection establishment phase, a customized E2E routing path is established by the SDN controller based on certain service requirement.

Therefore, an SDN-based connection establishment mechanism is designed to reduce both the amount of time consumed and the amount of signaling overhead for the connection establishment. A two-way handshake is applied to establish the connection for the forward and reverse directions respectively. The detailed procedure can be found in our previous work [23].

4.2.4.2 Data Transmission

SDATP Packet Format To achieve efficient slice-level data transmission, new packet formats are designed. Compared with the conventional TCP/IP, the header format is simplified in SDATP, with the help of SDN-based in-network transmission control. For example, the Acknowledgment field in the conventional TCP header used to acknowledge each received packet is removed from the SDATP packet header, which is replaced by the receiver-triggered packet loss detection. The SDATP packet header includes 24-byte required fields and 20-byte optional fields, as shown in Fig. 4.2, where slice ID consists of source and destination edge switch IP addresses and a pair of port numbers for E2E service delivery.

For packet forwarding, the required matching fields (i.e., slice ID) are extracted by the OpenFlow-switches, used to match the cached flow entries to make routing decisions. As some customized functionalities are introduced in SDATP, the header format is designed to support these new functionalities, such as in-path caching and retransmission, and caching-based congestion control. Therefore, in an SDATP packet header, the Flag and Optional fields are used for packet differentiation, including data packet, retransmission request (RR) packet, retransmission data (RD)

	1-8 bits	9-16 bits	17-24 bits	25-32 bits	
1	Protocol	Total length		Data offset	
2	Checksum		Flag		
3	Edge switch source IP address				
4	Edge switch destination IP address				Slice ID
5	Edge switch source port		Edge switch destination port		
6	Edge switch destination address				
7	Optional				

Fig. 4.2 The SDATP packet header format

packet, and CN packet. Those types of packets are to support the enhanced protocol functionalities.

Header Conversion/Reversion For compatibility with end hosts, the conventional TCP is applied between an end host and an edge switch. With the SDATP protocol for packet transmission between edge switches, packet header conversion and reversion for E2E communication are required, such as through the tunneling technology [24]. When a TCP packet is sent from the source node to the ingress edge switch, it is converted to an SDATP data packet, by adding a new SDATP header over the TCP header. When the SDATP packet arrives at the egress edge switch, it is reverted to a TCP packet by removing the SDATP header.

4.2.4.3　Packet Retransmission

We present the details of our proposed packet retransmission scheme, including in-path receiver-based packet loss detection and caching-based packet retransmission.

Receiver-Based Packet Loss Detection With activated retransmission functions, the in-network switches enable in-path receiver-based packet loss detection for fast packet loss recovery. If no packet loss occurs, packets are expected to be received in sequence following the linear network topology. However, packet loss leads to prolonged time interval for consecutive packet reception or a series of disordered packet reception. Accordingly, the retransmission node measures the time intervals for consecutive packet receptions and extracts the number of received disordered packets. Two thresholds, named *expected interarrival time* and *interarrival counter threshold*, are recorded at the retransmission node for packet loss detection. As loss can happen on RR and RD packets, a retransmission node should be capable of detecting the retransmitted packet loss and resending the RR packet. The detection is realized based on the measurement of each sampled packet retransmission delay, where the sample mean is estimated and used as the retransmission timeout threshold. The detailed information of how the loss detection thresholds are obtained is given in Appendix F. After packet loss is detected, the retransmission node sends an RR packet upstream to its preceding caching node(s) in its CR segment. The sequence number information used for identifying loss packet(s) is included in

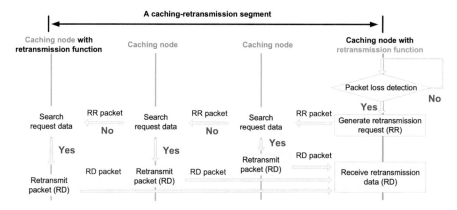

Fig. 4.3 An illustration of functionalities in a CR segment

the RR packet. This information is maintained by the retransmission node through establishing and updating an *expected packet list*, including which packets are not received and how the packets are detected to be lost. In addition, a *content window list* is established and maintained to record the received packets, which is used to update the expected packet list. More details of how to update these two lists are provided in Appendix G.

RD Packet Retransmissions by Caching Nodes The procedure of how caching and retransmission functions are operated within one CR segment is shown in Fig. 4.3. For packet loss recovery, both the first and last nodes in one CR segment are equipped with caching and retransmission functions. After an RR packet is received by a caching node, a range of sequence numbers and the retransmission triggering condition for the requested packets can be obtained. Based on the sequence number information, the caching node searches from its data caching buffer. If the requested packets are found, the RD packets are sent out downstream. Each RD packet also includes the timestamp from its received RR packet, which can be used for calculating packet retransmission delay. If current caching node cannot find the requested packets, the RR packet is forwarded upstream to preceding caching nodes until the requested packets are found.

4.2.4.4   Slice-Level Congestion Control

For each network slice, the transport-layer protocol operated between an end host and an edge switch follows TCP, while the proposed SDATP is adopted for the communications between two edge switches. Multiple slices share communication links and switches. With busy traffic flows, congestion happens when resource utilization is high, which leads to packet loss. Thus, we use packet loss as the indicator for congestion detection in the core network.

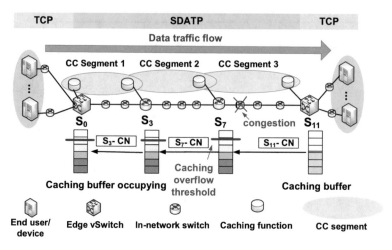

Fig. 4.4 Segment-level congestion control with $M = 11$

Caching-to-Caching Congestion Control For solving the link congestion problem, we introduce caching-to-caching congestion control between two edge switches to mitigate congestion as shown in Fig. 4.4, where only caching nodes are specified for illustration purpose. We define *one caching-caching (CC) network segment* which includes two consecutive caching nodes and the network switches/transmission links between them. Congestion detection for the nth ($n = 1, \ldots, N - 1$) CC segment is based on CN packets sent from downstream caching node $S_{C_{n+1}}$ to upstream caching node S_{C_n}. From the caching release window field of a CN packet, the upstream caching node S_{C_n} obtains the number of successfully received packets at caching node $S_{C_{n+1}}$ during last T. Since S_{C_n} determines how many packets are transmitted during last T, the number of lost packets over the nth CC segment, denoted by U_n, is calculated. Note that U_n is updated every T, and a large value of U_n indicates high congestion over the nth CC segment. Therefore, based on the information in a CN packet, link congestion can be detected.

We provide an example of caching-to-caching congestion control is given in Fig. 4.4, where 4 caching nodes (i.e., S_0, S_3, S_7, and S_{11}) are activated, and congestion happens in the third CC segment (i.e., between S_7 and S_{11}). Caching node S_7 detects the congestion from receiving a CN packet that indicates a high value of U_3, and starts to lower its sending rate to alleviate the congestion condition. Therefore, the local response time to congestion is T. The remaining caching buffer space of S_7 is decreased because of its lowered sending rate and the reduced number of released cached packets. Based on CN packets sent by node S_7, node S_3 obtains the remaining caching space of S_7 which is compared with a threshold. If the remaining space is lower than the threshold, node S_3 estimates that caching buffer overflow likely happens at S_7, and slows down its sending rate to lower the risk. The congestion information is spread out upstream until it reaches the ingress edge switch (S_0). After S_0 estimates the potential caching buffer overflow at S_3, it reduces the sending rate, and updates the rate of replying ACKs to the source node according to its remaining caching buffer size.

When congestion happens, the nearby in-network caching node detects congestion by observing packet loss. Then, preceding caching nodes spread out the information upstream to the source node by estimating the risk of caching buffer overflow, guided by a threshold. Thus, the reaction time to the congestion depends on the caching overflow threshold. With a smaller caching overflow threshold, each preceding caching node can achieve a faster overflow estimation, which gives the source node a faster reaction. On the contrary, with a larger caching overflow threshold, the source node will send out more packets at the premise of not aggravating the congestion. By reducing the waiting time at the source node, a lower E2E transmission delay can be achieved. Thus, the caching overflow threshold is an important design parameter, due to its impact on both congestion reaction time of the source node and E2E transmission delay.

4.2.5 Optimized Probabilistic Caching Policy

To satisfy high-reliability and low-latency service requirements, we introduce in-path caching-based retransmission in SDATP, which is achieved through activating caching and retransmission functions at in-network switches. Furthermore, to minimize the average number of packet retransmission hops, we propose an optimized packet caching policy, to determine caching node placement, packet caching probability, and the number of activated caching nodes.

With the embedded network topology supporting the time-critical service shown in Fig. 4.1, a packet loss at the mth link indicates that the packet has been successfully transmitted through the first $(m - 1)$ links and lost at the mth link (or transmission hop), the probability of which is given by

$$
\begin{cases}
p_1, & \text{if } m = 1 \\
Q_s(m - 1) \times q_m, & \text{if } m = 2, 3, \dots, M
\end{cases}
\tag{4.2}
$$

where q_m denotes the packet loss probability over link m ($m = 1, 2, \dots, M$) between edge switches caused by network congestion and/or link errors, $Q_s(m - 1) = \prod_{i=1}^{m-1} (1 - p_i)$ is the successful transmission probability of a packet over the first $(m - 1)$ transmission hops before arriving at switch S_m. Thus, the probability of packet loss between S_m and edge switch S_M is $[Q_s(m) - Q_s(M)]$. Within a time interval of T, switch S_m successfully receives $Y \cdot Q_s(m)$ packets, among which $Y \cdot [Q_s(m) - Q_s(M)]$ packets are lost on average between S_m and S_M.

For packet retransmission, both the delay and transmission resource consumption are proportional to the required retransmission hops (RRHs). Thus, we describe the performance of in-path caching-based retransmission in terms of RRH number in one T. Specifically, the benefit achieved by exploiting in-path caching nodes is the eliminated RRHs compared with no in-path caching nodes. To improve the caching efficiency, packets can only be cached once, indicating the already cached packets cannot be cached again. Thus, the probability of a packet cached at S_{C_n} after passing through a sequence of caching nodes $\{S_{C_i}, i = 1, 2, \ldots, n - 1\}$ is given by

$$
P_t(C_n) = \begin{cases} P_c(C_n), & \text{if } n = 1 \\ \prod_{i=1}^{n-1} [1 - P_c(C_i)] \cdot P_c(C_n), & \text{if } n = 2, 3, \ldots, N \end{cases} \tag{4.3}
$$

where $P_c(C_n)$ denotes the caching probability of a packet passing through caching node S_{C_n}.

With in-path caching, if a packet cached at S_{C_n} is lost and is requested for retransmission, it can be retransmitted by caching node S_{C_n}. Otherwise, it needs to be retransmitted from the ingress edge switch. Compared to the case without in-path caching, at least C_n transmission hops can be avoided for each packet retransmission. With caching buffer capacity B_{C_n} (i.e., maximum number of cached packets) allocated to S_{C_n}, the number of packets that can be cached at S_{C_n} is $\min\left[Y \cdot Q_s(C_n) \cdot P_t(C_n), B_{C_n}\right]$. Then, the reduction in the average number of RRHs for packets cached at S_{C_n} during T is given by

$$
H(C_n) = \min\left[Y Q_s(C_n) P_t(C_n), B_{C_n}\right] \frac{Q_s(C_n) - Q_s(M)}{Q_s(C_n)} C_n. \tag{4.4}
$$

4.2.5.1　Problem Formulation

The performance gain is represented by a ratio of the total number of eliminated RRHs for packets cached at all caching nodes to the total number of RRHs for all lost packets without in-path caching. That is,

$$
\frac{\sum\limits_{n=1}^{N} H(C_n)}{Y \cdot [1 - Q_s(M)] \cdot M} = \frac{\sum\limits_{n=1}^{N} \min\left[Y \cdot Q_s(C_n) \cdot P_t(C_n), B_{C_n}\right] \frac{Q_s(C_n) - Q_s(M)}{Q_s(C_n)} \cdot C_n}{Y \cdot [1 - Q_s(M)] \cdot M}. \tag{4.5}
$$

Since Y is a random variable, we calculate the expectation of (4.5) as

$$
\begin{aligned}
G(N) &= \mathbb{E}\left\{\frac{\sum_{n=1}^{N} H(C_n)}{Y \cdot [1 - Q_s(M)] \cdot M}\right\} \\
&= \frac{\sum_{n=1}^{N} \frac{Q_s(C_n) - Q_s(M)}{Q_s(C_n)} C_n \mathbb{E}\left\{\min\left[Q_s(C_n) P_t(C_n), \frac{B_{C_n}}{Y}\right]\right\}}{[1 - Q_s(M)] \cdot M}
\end{aligned} \tag{4.6}
$$

where $\mathbb{E}\{\cdot\}$ is the expectation operation. In (4.6), we have

$$
\begin{aligned}
&\mathbb{E}\left\{\min\left[Q_s(C_n)\cdot P_t(C_n),\frac{B_{C_n}}{Y}\right]\right\}\\
&= F(C_n)\cdot Q_s(C_n)\cdot P_t(C_n) + B_{C_n}\cdot\sum_{k=R(C_n)+1}^{\infty}\frac{1}{k}\cdot\frac{\lambda^k}{k!}e^{-\lambda}
\end{aligned}
\tag{4.7}
$$

where $F(C_n)$ denotes the caching buffer non-overflow probability (i.e., caching reliability) at switch S_{C_n}, given by

$$
F(C_n) = Pr\left\{Y Q_s(C_n) P_t(C_n) \le B_{C_n}\right\} = Pr\left\{Y \le R(C_n)\right\}.
\tag{4.8}
$$

For Y is limited by the caching buffer capacity of S_{C_n}, the traffic rate upper bound, $R(C_n)$ is given by

$$
R(C_n) = \left\lfloor\frac{B_{C_n}}{Q_s(C_n)\cdot P_t(C_n)}\right\rfloor
\tag{4.9}
$$

where $\lfloor\cdot\rfloor$ is the floor function.

It is indicated in (4.5) that the performance gain depends on the in-path caching policy. Therefore, to minimize the average packet retransmission delay (i.e., to maximize $G(N)$), we consider to optimize the number of enabled caching nodes, N, caching function placement indicated by $\mathbb{C}(N) = \{C_1, C_2, \ldots, C_N\}$, and the set of packet caching probabilities $\mathbb{P}(N) = \{P_t(C_1), P_t(C_2), \ldots, P_t(C_N)\}$. The optimization problem for probabilistic caching is formulated to maximize $G(N)$ as

$$
\textbf{(P1)}: \max_{N,\mathbb{C}(N),\mathbb{P}(N)} G(N)
$$

s.t. $\begin{cases} 1 \le N \le M-1, N \in \mathbf{Z}^+ & (4.10a) \\ 1 \le C_1 < C_2 < \ldots < C_N \le M-1 & (4.10b) \\ C_n \in \mathbf{Z}^+, n = 1, 2, \ldots, N & (4.10c) \\ 0 < P_c(C_n) \le 1, n = 1, 2, \ldots, N. & (4.10d) \end{cases}$

This optimization problem is an MINLP problem and is difficult to solve due to high computational complexity. In the following, based on design principles for $\mathbb{P}(N)$, we derive $\mathbb{P}(N)$ in terms of N and $\mathbb{C}(N)$, upon which **(P1)** can be simplified with a reduced number of decision variables.

There are two principles for determining $\mathbb{P}(N)$: (1) All transmitted packets should be cached in path between the edge switches (i.e., $P_c(C_N) = 1$ and $\sum_{n=1}^{N} P_t(C_n) = 1$), which guarantees that all lost packets can be retransmitted from in-path caching nodes; (2) The caching buffer overflow probabilities at all caching-enabled switches should be equal, in order to balance the caching resource utilization at different in-path caching nodes. To satisfy principle (2), $R(C_n)$ (or $F(C_n)$) is the same for all $n \in \{1, 2, \ldots, N\}$. Thus, we use R_N and F_N to represent $R(C_n)$ and $F(C_n)$ for simplicity. Based on (4.9), $P_t(C_n)$ is given by

$$P_t(C_n) = \frac{B_{C_n}}{Q_s(C_n) \cdot R_N} \tag{4.11}$$

and the summation of $P_t(C_n)$ over all caching nodes is

$$\sum_{n=1}^{N} P_t(C_n) = \sum_{n=1}^{N} \frac{B_{C_n}}{Q_s(C_n) \cdot R_N} = \frac{1}{R_N} \cdot \sum_{n=1}^{N} \frac{B_{C_n}}{Q_s(C_n)} = 1. \tag{4.12}$$

Therefore, R_N is expressed as

$$R_N = \sum_{n=1}^{N} \frac{B_{C_n}}{Q_s(C_n)} \tag{4.13}$$

and F_N is calculated based on (4.13) as

$$F_N = Pr\{Y \leq R_N\} = \sum_{k=0}^{R_N} \frac{\lambda^k}{k!} e^{-\lambda}. \tag{4.14}$$

Based on (4.11) and (4.13), $\mathbb{P}(N)$ is expressed as a function of N and $\mathbb{C}(N)$, and (**P1**) is simplified in terms of decision variables.

To further reduce the problem complexity, we simplify the objective function of (**P1**) through algebraic manipulation. In (**P1**), $G(N)$ is a summation of $G_r(N)$ and $G_o(N)$, which represents a combination of performance gains in caching buffer non-overflow and overflow cases, respectively. Thus, we obtain $G_r(N)$ and $G_o(N)$ as

$$G_r(N) = \frac{\sum_{n=1}^{N} [Q_s(C_n) - Q_s(M)] \cdot C_n \cdot \frac{B_{C_n}}{Q_s(C_n)}}{[1 - Q_s(M)] \cdot M \cdot \sum_{n=1}^{N} \frac{B_{C_n}}{Q_s(C_n)}} \cdot F_N \tag{4.15}$$

and

$$G_o(N) = \frac{\sum_{n=1}^{N} [Q_s(C_n) - Q_s(M)] \cdot C_n \cdot \frac{B_{C_n}}{Q_s(C_n)} \cdot A}{[1 - Q_s(M)] \cdot M} \tag{4.16}$$

where $A = \sum_{k=R_N+1}^{\infty} \frac{1}{k} \cdot \frac{\lambda^k}{k!} e^{-\lambda}$. Since the calculation of infinite series in A incurs high computational complexity, we calculate its lower and upper bounds, denoted by A_l and A_u. We first determine A_l as

$$A_l = \sum_{k=R_N+1}^{\infty} \frac{1}{k+1} \cdot \frac{\lambda^k}{k!} e^{-\lambda} = \frac{1}{\lambda} \sum_{k=R_N+2}^{\infty} \frac{\lambda^k}{k!} e^{-\lambda}$$

$$= \frac{1}{\lambda} \left(1 - F_N - Pr\{Y = R_N + 1\} \right).$$

(4.17)

Then, we have

$$A - A_l = \sum_{k=R_N+1}^{\infty} \left(\frac{1}{k} - \frac{1}{k+1} \right) \cdot \frac{\lambda^k}{k!} e^{-\lambda}$$

$$< \frac{1}{\lambda} \cdot \frac{1}{R_N+1} \sum_{k=R_N+2}^{\infty} \frac{\lambda^k}{k!} e^{-\lambda}$$

(4.18)

$$= \frac{A_l}{R_N+1}.$$

From (4.18), the upper bound of A is derived as

$$A < A_l + \frac{A_l}{R_N+1} = A_l \cdot \frac{R_N+2}{R_N+1} = A_u.$$

(4.19)

With A_l, the lower bound of $G_o(N)$ is expressed as

$$G_{o,l}(N) = \frac{\sum_{n=1}^{N} [Q_s(C_n) - Q_s(M)] \cdot C_n \cdot \frac{B_{C_n}}{Q_s(C_n)}}{[1 - Q_s(M)] \cdot M \cdot \lambda}$$

$$\cdot (1 - F_N - Pr\{Y = R_N + 1\}).$$

(4.20)

Using A_u, the upper bound of $G_o(N)$ is calculated as

$$G_{o,u}(N) = \frac{R_N+2}{R_N+1} \cdot G_{o,l}(N).$$

(4.21)

When the caching buffer overflow happens at S_{C_n}, some cached packets are dropped from the buffer. Later on, if the retransmission requests are triggered for those cached packets, the packets need to be retransmitted from the ingress edge switch, which requires C_n hops more than that if retransmitted from S_{C_n}. Also, the caching resources for those overflowed packets at S_{C_n} are wasted without any performance gain. Since caching buffer overflow leads to a decrease of performance gain and a wastage of caching resources, the caching buffer non-overflow probability F_N

for the optimal caching strategy tends to 1. Hence, R_N is expected not to be less than λ, which is the median of Y as discussed in Sect. 4.2.3. With λ ranging from tens to hundreds of packets per T, the gap between $G_{o,u}(N)$ and $G_{o,l}(N)$ is small. Therefore, we use $G_{o,l}(N)$ to estimate $G_o(N)$. Then, the objective function $G(N)$ in (**P1**) is simplified to $G_r(N) + G_{o,l}(N)$, and (**P1**) is transformed to

$$(\mathbf{P2}): \max_{N,\mathbb{C}(N)} \ G_r(N) + G_{o,l}(N)$$

$$\text{s.t.} \begin{cases} 1 \le N \le M - 1, N \in \mathbf{Z}^+ & \text{(4.22a)} \\ 1 \le C_1 < C_2 < \ldots < C_N \le M - 1 & \text{(4.22b)} \\ C_n \in \mathbf{Z}^+, n = 1, 2, \ldots, N. & \text{(4.22c)} \end{cases}$$

4.2.5.2 Optimized Probabilistic Caching

By solving (**P2**), an optimal number of caching nodes, N, and its corresponding $\mathbb{C}(N)$ can be obtained to maximize $G(N)$. However, (**P2**) is a nonlinear integer programming problem, which is difficult to solve [25]. For tractability, we design a low-complexity heuristic algorithm as Algorithm 3 to jointly optimize N and $\mathbb{C}(N)$.

If a caching node is placed near the receiving node, a large number of retransmission hops (C_n) is avoided for each individual retransmitted packet, but the cached packets experience a low caching efficiency (i.e., the probability of being requested for retransmission). Therefore, for caching placement $\mathbb{C}(N)$, there is a trade-off between the reduced retransmission hops and the caching efficiency. In terms of the number of caching nodes, a large value of N means more caching resources are available to guarantee a low caching buffer overflow probability. However, packets are distributively cached at caching nodes, with decreased caching resource utilization. A maximal $G(N)$ can be achieved through balancing the trade-off between the buffer overflow probability and the utilization of caching resources of switches.

Specifically, we define caching weight W_m for in-network switch S_m as the product of its caching efficiency, reduced retransmission hops of one cached packet (m), and caching buffer size (B_m) to represent its contribution to the overall performance gain. That is

$$W_m = \frac{[Q_s(m) - Q_s(M)]}{Q_s(m)} \cdot m \cdot B_m \tag{4.23}$$

where the caching efficiency of S_m is given by $\frac{[Q_s(m) - Q_s(M)]}{Q_s(m)}$. The higher the value of W_m, the greater the contribution. Therefore, the caching nodes are selected based on W_m in our proposed probabilistic caching algorithm (Algorithm 3). Given N, $\mathbb{C}(N)$ is determined in line 5 and line 6 of Algorithm 3, and $G(N)$ is calculated

---

Algorithm 3: The probabilistic caching algorithm

Require: Loss probability $\{q_1, q_2, \ldots, q_M\}$, caching buffer resource $\{B_1, B_2, \ldots, B_{M-1}\}$;
  Traffic load information, packet arrival rate λ
Ensure: Caching placement set $\mathbb{C}(N^*)$ and caching probability set $\mathbb{P}(N^*)$
  1: Initialization: Set performance gain G^* to 0
  2: Calculate W_m for each in-network switch
  3: Rank W_m in a descending order
  4: **for** $N = 1 : M - 1$ **do**
  5:   Find N switches with largest W_m
  6:   Get the index set as caching node set $\mathbb{C}(N)$
  7:   Calculate $G(N)$ for $\mathbb{C}(N)$ based on (4.15) and (4.20)
  8:   **if** $G(N) > G^*$ **then**
  9:     Set N^* to N
 10:     Set $\mathbb{C}(N^*)$ to $\mathbb{C}(N)$
 11:     Set G^* to $G(N)$
 12:   **end if**
 13: **end for**
 14: Calculate caching probability $\mathbb{P}(N^*)$ with N^* and $\mathbb{C}(N^*)$ based on (4.11) and (4.13)

---

based on (4.15) and (4.20). Then, we iterate N to find the N value that achieves the maximal $G(N)$ (from line 8 to line 12).

4.2.5.3 Time Complexity

The time complexity for calculating $\left[G_r(N) + G_{o,l}(N)\right]$ in **(P2)**, which linearly increases with the number of caching nodes N, i.e., $O(N)$. Based on Algorithm 3, $\left[G_r(N) + G_{o,l}(N)\right]$ in **(P2)** is calculated for each N ($N = 1, 2, \ldots, M - 1$). Thus, the time complexity is given by

$$\sum_{N=1}^{M-1} O(N) = O\left(\sum_{N=1}^{M-1} N\right) = O\left(M^2\right). \tag{4.24}$$

For the brute-force method, there are C_{M-1}^N candidate sets for caching placement for a given N value, and $\left[G_r(N) + G_{o,l}(N)\right]$ needs to be calculated for each candidate set. Therefore, the total time complexity is derived as

$$\sum_{N=1}^{M-1} C_{M-1}^N \cdot O(N) = O\left(\sum_{N=1}^{M-1} \frac{(M-1)!}{N!(M-1-N)!} \cdot N\right) = O\left(M \cdot 2^M\right). \tag{4.25}$$

Compared with the brute-force method, our proposed probabilistic caching policy generates a solution with a much lower time complexity.

4.2.6 Numerical Results

In this subsection, both analytical and simulation results are presented to demonstrate the effectiveness of the proposed SDATP. The SDN/NFV-based network architecture is shown in Fig. 4.5, in which two separated virtual machines, with 8 GB physical memory and 12 virtual processors, are utilized to simulate the data and control planes, respectively. In the data plane, the network elements are emulated by Mininet, including end hosts, edge/in-network switches, and transmission links [26]. Both edge and in-network switches are OpenFlow vSwitch [27]. In the control plane, the SDN controller is implemented by the Ryu framework [28]. For the SDN southbound interface, we use OpenFlow version 1.3.0 [29] for simulation.

An SDN/NFV-based linear network topology is deployed for supporting a time-critical service flow, as shown in Fig. 4.6, where two E2E source-destination node pairs, $(A.1, B.1)$ and $(A.2, B.2)$, are connected via eleven switches (two edge switches and nine core network switches) under the SDN control. Traffic flows from both source nodes are aggregated at edge switch S_0 and are then sent to corresponding destination nodes through egress edge switch S_{10}. A detailed simulation setting is given in Table 4.1. The caching function is activated at ingress edge switch, S_0, while both caching and retransmission functions are activated at egress edge switch, S_{10}. In addition, in-path functions are activated based on the proposed caching placement scheme in order to adapt to varying network conditions.

Fig. 4.5 Simulation platform for the proposed protocol to support a time-critical service

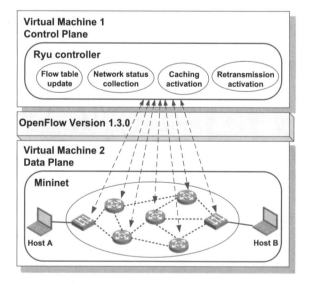

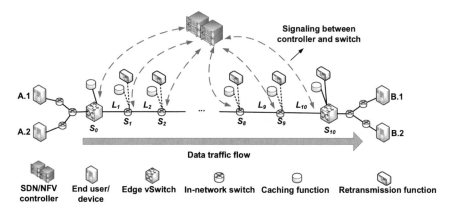

Fig. 4.6 Virtual network topology for supporting a time-critical service

Table 4.1 Simulation parameters

Packet size	1400 byte	
Between user and edge node		
Source node	Data sending rate	1.25 Mbps
Link	Transmission rate	2.5 Mbps
	Propagation delay	10 ms
Between edge nodes		
Switch	Transmission data buffer	100 packets
	Caching buffer	5–20 packets
Link	Transmission rate	5 Mbps
	Propagation delay	10 ms
	Packet loss rate	0.05%

4.2.6.1 Adaptive Probabilistic Caching Policy

We evaluate the adaptiveness of the proposed probabilistic caching policy with respect to the varying packet arrival rate. Analytical results are obtained using MATLAB, where link packet loss rates and switch caching buffer sizes are set using parameters in Table 4.1.

The number of active devices (e.g., activating devices in smart alarming system) varies with time, leading to different traffic volumes during peak and off-peak hours. To evaluate the adaptiveness of probabilistic caching policy, we vary the packet arrival rate of an aggregated flow, from 10 packets per T to 90 packets per T. The performance gain defined in (4.6) indicates the improvement of using in-path caching on the reduced number of retransmission hops. Given packet arrival rate, the caching policy (i.e., caching placement and caching probability) is shown in Fig. 4.7a. The caching nodes with normalized weight W_m ($m = 1, 2 \ldots, M - 1$) are differentiated by colours as shown in Fig. 4.7b. The height of each coloured bar in Fig. 4.7a denotes the caching probability (i.e., $P_t(C_n)$ in (4.3)) of a specific node. To ensure all transmitted packets are cached in path, the summation of caching

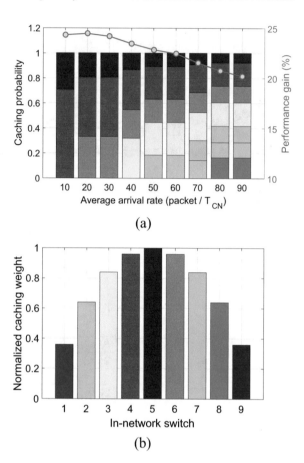

Fig. 4.7 Adaptiveness to the packet arrival rate: (a) caching node placement and performance gain; (b) normalized caching weight for each node

probabilities equals to 1. It shows that more caching nodes are activated upon a traffic load increase, which is necessary to guarantee caching reliability. However, to accommodate the increasing traffic, some nodes with a small weight are selected, which leads to a decreased performance gain.

We also compare the proposed policy with two other in-path caching policies, called *RP* and *HP* policies. In RP policy, the in-path caching nodes are randomly selected. In HP policy, the nodes immediately before the links with high packet loss probability are activated as in-path caching nodes. Given the packet arrival rate is 50 packets per T, caching placement and packet caching probability for each node are designed based on the proposed probabilistic caching (PC) algorithm, RP, and HP methods. In addition, we compare the PC method with a modified PC (*MPC*), in which each packet can be cached at multiple locations. For the proposed PC scheme, the number of activated caching nodes is 4, and nearly 24% of retransmission hops can be avoided as shown in Fig. 4.8, which outperforms the other three methods. In addition, since PC outperforms MPC, it is reasonable that we make each packet only

Fig. 4.8 Performance gain
achieved by different methods

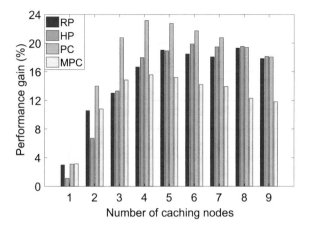

Fig. 4.9 Performance gain
versus number of caching
nodes

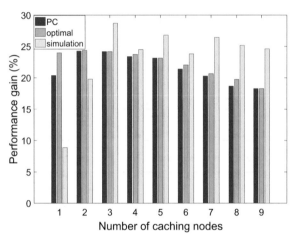

be cached once over the network path between the edge switches in the simulated
networking environment.

To evaluate the accuracy of proposed algorithm, we compare the performance
gain of the PC policy and the optimal policy achieved by brute-force algorithm
shown in Fig. 4.9, where packet arrival rate is 30 packets per T. It shows that
the optimal number N^* of activated caching nodes exists and the gap of optimal
performance gains between the PC policy and the brute-force method is small. We
also demonstrate the performance gain through simulations. It can be seen that the
simulation results closely match the analytical results under different number of
caching nodes.

4.2.6.2   E2E Packet Delay

The E2E packet delay of our proposed SDATP is compared with that of TCP. We compare the average E2E delay (i.e., the duration from the time instant that a packet is sent from the source node till the instant that it is received by the destination node, averaged over all received packets) between SDATP and TCP for packets sent from $A.1$ to $B.1$. Two scenarios are considered: (a) Packet loss happens due to link errors; (b) Packet loss happens due to congestion. Based on the link propagation delay setting, the E2E delay for packet transmission, assuming no packet loss happens, is 120 ms, which is the lower bound of the E2E packet delay. Due to time consumed in packet loss detection and packet retransmission, the average E2E delay in both scenarios are larger than 120 ms.

(1) Packet Loss Due to Link Errors Figure 4.10 shows the average E2E delay of TCP and SDATP when the E2E packet loss rate varies, where packet loss rate between a user and an edge switch ranges from 0% to 1% and between in-network switches is 0.05%.

It is observed that SDATP achieves a shorter average E2E delay than that of TCP. For TCP, packet loss is detected at the source, with an E2E round trip packet transmission time, before a packet retransmission is performed. For SDATP, the in-path caching-based retransmission enables packet loss detection within a CR segment. The time duration for loss detection and retransmission triggering is much shorter than the E2E round trip time, leading to a shorter average E2E delay.

(2) Packet Loss Due to Congestion Consider that packet loss over all links in the network are due to congestion. The transmission rate over the physical link between S_5 and S_6 is denoted by C Mbps. We set C smaller than 5 Mbps such that congestion can happen due to transmission bottleneck between S_5 and S_6.

Figure 4.11 shows a comparison of the average E2E delay between TCP and SDATP, where C varies from 0.5 to 3 Mbps to indicate different levels of congestion.

Fig. 4.10 A comparison of average E2E packet delay between TCP and SDATP without congestion

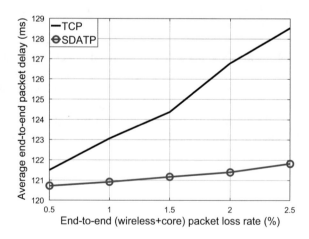

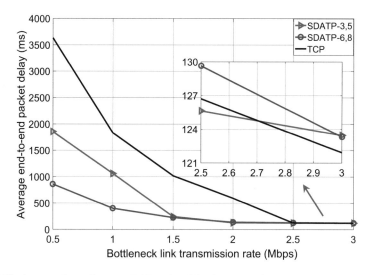

Fig. 4.11   A comparison of average E2E packet delay between TCP and SDATP with congestion

For SDATP, the locations of congestion and caching nodes have impact on both congestion control and retransmission performance. Therefore, we place the caching nodes before (i.e., at S_3 and S_5) and after (i.e., at S_6 and S_8) the congestion link, respectively. When C varies from 0.5 to 2 Mbps, the congestion condition is severe, and packets are lost frequently due to data buffer overflow. Compared with TCP, the average E2E packet delay is reduced for SDATP due to early packet loss detection and shortened packet retransmission delay, as shown in Fig. 4.11. Moreover, placing caching nodes after the congestion link outperforms placing before the congestion, since the congestion is detected earlier by S_0, with CN packets sent from S_6. However, when the congestion level is low (i.e., C varies from 2.5 to 3 Mbps), SDATP has a higher average E2E packet delay than that of TCP, due to time consumption for packet caching in SDATP. In contrast to heavy congestion, placing caching nodes before the congestion link acheives better performance, due to fast packet retransmission from S_5.

4.3   Protocol Customization for Video-Streaming Services

As discussed in Sect. 4.2, with SDN and NFV, fine-grained transmission control functionalities can be realized in network to enforce more efficient protocol operation to achieve service-oriented protocol customization. In a 5G E2E network architecture, edge nodes (i.e., ingress and egress nodes), which connect end users to a core network, are augmented with higher protocol-layer functionalities, e.g., transmission control, user data logger, and mobility management, to enable more delicate control for each sliced virtual network. In this section, a transmission protocol operated at edge nodes of a 5G core network is customized for supporting a VoD streaming service, where in-network selective caching and enhanced transmission functionalities are enabled. Specifically, to mitigate the network congestion level, the ingress node caches certain number of video packets through selective caching functionality to reduce the E2E delay by taking into consideration the video traffic load and the available resources along the E2E transmission path. The prediction of the number of video packet arrivals in each time slot at the ingress node is based on the autoregressive integrated moving average (ARIMA) model for making proactive packet caching decisions. To enhance E2E throughput without further incurring new congestion events, the enhanced transmission functionality is activated by re-sending some of the cached video packets from the ingress node to the video clients when the network condition improves. To capture the implicit relationship between congestion control and QoS performance with unknown video traffic arrival statistics, an action selection module based on the multi-armed bandit (MAB) framework is employed to select proper transmission control actions at the ingress node via balancing exploration and exploitation. The action-selection strategy is updated by observing the feedback reward at the end of each time slot. The *cold-start* problem exists in the considered scenario when the protocol operates under new network conditions [30]. By taking into consideration the cold-start issue, we formulate the control action selection problem as a contextual bandit problem [30, 31]. The LinUCB algorithm is adopted to solve the formulated problem (i.e., determine the control action in each time slot), which has been theoretically proved to have strong regret bound [31].

4.3.1 Network Model

Consider a 5G core network where traffic of certain service type from different end source nodes is aggregated as one traffic flow at a core network ingress node. As shown in Fig. 4.12, multiple traffic flows traverse the core network. Each traffic flow is required to be processed by a chain of VNFs which are implemented on a set of NFV nodes. Between consecutive NFV nodes, there are a number of in-network switches connected by physical links to forward the traffic. The transmission path of each traffic flow in the core network is determined by the SDN controller [32]. To improve resource utilization, more than one traffic flow often passes a common set of network elements (in-network switches, physical links, or NFV nodes) and share the same pool of physical resources [19]. Two types of resources are considered, i.e., (1) computing resources at NFV nodes, and (2) transmission resources over physical links [15]. Given the transmission path and the allocated resources of a traffic flow, a customized transmission protocol is employed to achieve service-oriented control. To ensure QoS isolation among different services, a network slice is created to support the packet transmission of each flow, which consists of an E2E

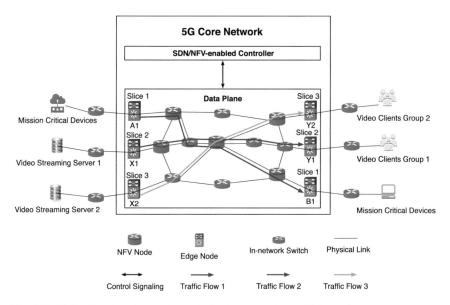

Fig. 4.12 Embedded virtual network topologies for supporting multiple services

transmission path with properly allocated resources and customized transmission protocol.

A unicast VoD streaming slice has a linear topology between a pair of edge nodes (e.g., Slice 2 in Fig. 4.12). The set of nodes in the slice is denoted by $\mathcal{V} = \{V_1, V_2, \ldots, V_L\}$, where L is the total number of nodes in the slice. A node is either an in-network switch or an NFV node which has a first-in-first-out (FIFO) buffer to queue arrived packets. We assume that the buffer always has sufficient space to queue a newly arrived packet. The bottleneck resource type of an in-network switch and NFV node is transmission resources and computing resources, respectively. Here, the resource type of a node in a VoD streaming slice refers to its bottleneck resource type. At each node, the video traffic flow shares the resources with multiple cross-traffic flows. The number of cross-traffic flows traversing node V_l is denoted by M_l ($l = 1, 2, \ldots, L$). Time is partitioned into slots of constant duration T_s [33]. Denote t_k as the time instant when the k-th time slot starts. At t_k, the average traffic rate of the j-th cross-traffic flow at V_l is calculated as [34]

$$\lambda_j^{(l)}(t_k) = \left\lfloor \frac{n_j(t_k) - n_j(t_k - T_s)}{T_s} \right\rfloor \tag{4.26}$$

where $n_j(t_k)$ represents the number of packets of the j-th flow that have arrived at V_l by t_k, $\lfloor \cdot \rfloor$ is the floor function. Denote by C_l the total capacity (in packet/s) of V_l. The available resources (in packet/s) on V_l at time t_k, denoted by $r_l(t_k)$, is given by [34]

$$r_l(t_k) = C_l - \sum_{j=1}^{M_l} \lambda_j^{(l)}(t_k). \tag{4.27}$$

The E2E available resources, $r(k)$, at the k-th time slot for a VoD streaming slice is determined as

$$r(k) = \min\{r_1(t_k), r_2(t_k), \ldots, r_L(t_k)\}. \tag{4.28}$$

The server-side edge node (client-side edge node) of the VoD streaming slice is the ingress node (egress node) which is assumed to have enough caching resources to buffer the packets chosen by the selective caching functionality. For example, nodes X_1 and Y_1 in Fig. 4.12 are the ingress and egress nodes, respectively, of Slice 2. For backward compatibility on end hosts, the ingress (egress) node is an in-network proxy server which maintains the TCP connections with the video server (clients) [35]. The ingress node replies an ACK packet to the video server for every received video packet. All the video packets received by the egress node are converted to TCP packets, which are copied and cached at the egress node, and are then forwarded to the corresponding video clients. Each video client replies an ACK packet of each received video packet for acknowledgement. When the egress node receives an ACK packet from a video client, it removes the corresponding video packet from the egress node caching buffer. However, if a video packet is lost between the egress node and the video client, the egress node either receives duplicate ACKs or triggers retransmission timeout. In this case, the egress node retransmits the lost packet and activates the TCP congestion control mechanism.

4.3.2   VoD Streaming System

The scalable video coding (SVC) technique is used to encode video files in the server [36]. Each video is divided into a series of video segments. Denote by Δ_s the length of a segment. Each segment is further encoded into several layers, including one base layer and N_e enhancement layers. Different layers of a video segment can be stored and streamed independently in form of small video chunks. The base-layer chunks are required for video segment decoding at clients. An enhancement-layer chunk can be decoded only if all the lower enhancement-layer chunks and the base-layer chunk from the same video segment are received by the client. The more enhancement-layer chunks are received, the higher video quality will be. Before sending the chunks into the network, each chunk is fragmented and encapsulated into multiple video packets. The quality of the streamed video segments, indicated by the number of SVC layers, is controlled by the video clients [37, 38]. When all the base-layer packets of the requested segments are received by a client, the client needs to determine the requested quality for the following several segments based on the current buffer level, i.e., the number of playable video segments in the client buffer. The desired quality information is transmitted to the video server by the HTTP GET message [39].

4.3.3 Protocol Functionalities

To achieve in-network control for a VoD streaming slice, the proposed transmission protocol incorporates the following functionalities: header conversion functionality, selective caching functionality, and enhanced transmission functionality. When a congestion event occurs in the VoD streaming slice, the ingress node selectively caches incoming packets into the caching buffer. Once the network condition improves, the packets which can enhance video quality are retrieved from the caching buffer for enhanced transmission. At the beginning of each time slot, the ingress node of VoD streaming slice selects appropriate functionality based on the network condition. The protocol functionality details are described in the following:

1. Header conversion functionality—It is deployed at the ingress node to add a header over all video packets passing through [39]. The header format is shown in Fig. 4.13. Between the edge nodes of a VoD streaming slice, the source (destination) IP address of the video packet is indicated by the *Ingress (Egress) Node Address* field. The sending (receiving) port number at the ingress (egress) node is included in the *Ingress (Egress) Node Port Number* field. The fields enclosed by the red dashed rectangular box is referred to as *slice ID* for slice differentiation. The *Protocol* field indicates the applied transmission protocol for the video traffic flows in the core network. The fields of *Total Length*, *Data Offset* and *Checksum* are necessary to packets traversing the network. The *Flag* field is used to differentiate the types of packets in the VoD streaming slice. The

	1 - 8 bits	9 - 16 bits	17 - 24 bits	25 – 32 bits
1	Protocol	Total Length		Data Offset
2	Checksum		Flag	
3	Ingress Node Address			
4	Egress Node Address			
5	Ingress Node Port Number		Egress Node Port Number	
6-8	Client ID			
9	Segment Number			Layer Number

Fig. 4.13 The proposed protocol packet header format

Client ID contains the IP addresses and port numbers of the server and clients. The *Segment Number* and *Layer Number* of a video packet are extracted from the application layer payload at the ingress node. Note that the layer number of base-layer packets and i-th enhancement-layer packets is 0 and i respectively;

2. Selective caching functionality—An SVC codec enables flexible video decoding, and video contents can be successfully decoded even in the absence of enhancement-layer packets. Hence, higher layer packets are selectively cached in the network, without significant degradation of user experience. By utilizing the caching resources, instead of dropping packets when network is congested, we design a selective in-network caching policy to temporarily store certain packets at the ingress node for a fast response to network congestions;

3. Enhanced transmission functionality—To increase the video quality once the network condition allows, we design an enhanced transmission protocol functionality. At each time slot when the enhanced transmission is activated, the ingress node determines how many cached packets should be transmitted in the slot, and the cached packets are pushed from the caching buffer to the VoD streaming slice.

4.3.4   Performance Metrics

To evaluate the performance of the proposed protocol, we consider the following four QoS metrics in VoD streaming systems:

1. Average E2E delay—the E2E delay, consisting of packet queueing delay, packet processing delay, link transmission delay and propagation delay experienced in the core network, averaged over all the packets passing through the egress node during a time slot;

2. Throughput—the total number of video packets from a VoD streaming slice passing through the egress node in 1 s;

3. Goodput ratio—the number of packets with bounded E2E delay over total number of packets passing through the egress node during one time slot;

4. Resource utilization—the throughput over E2E available resources for the VoD streaming slice.

4.3.5   Learning-Based Transmission Protocol

In this section, the transmission protocol customized for VoD streaming services is presented, which includes three main components: (1) traffic prediction module, (2) E2E available resource measurement module, and (3) action selection module for selecting control actions.

4.3.5.1 Protocol Framework

The proposed protocol controls the packet queueing delay during the network congestion and enhances the throughput once the congestion event is over by adjusting the traffic load for a VoD streaming slice. It achieves traffic management by taking different control actions at the ingress node. When the selective caching functionality is activated, some incoming video packets from the video traffic flow are cached in the caching buffer at the ingress node. If the enhanced transmission functionality is enabled, the cached video packets are transmitted from the ingress node to the video clients. For protocol operation, three functional modules are implemented at the ingress node of a VoD streaming slice, i.e., video traffic prediction module, E2E available resources measurement module, and action selection module. The relationship among the modules is shown in Fig. 4.14. The video traffic prediction module estimates the traffic load of the next time slot based on the traffic loads observed in the last several time slots. The E2E available resources measurement module is used to monitor the available resources for a VoD streaming slice during the network operation. The action selection module is the key of the proposed protocol which selects the control action in each time slot based on the output of the other two functional modules.

Denote by $\hat{t}(k)$ the output of video traffic prediction module. To control the dimensionality of the action space, the selective caching and enhanced transmission functionalities operate at SVC layer level and packet chunk level, respectively. The packet chunks for enhanced transmission are labeled as ET-chunks. All the ET-chunks contain the same number, N_c, of video packets. Denote by N_E the predetermined maximum number of ET-chunks transmitted in one time slot. Let $\mathcal{A}$ denote the set of all possible control actions, each of which is denoted as a two-element tuple, $(i, j) \in \mathcal{A}$, where $i = 0, 1, \ldots, N_e$ and $j = 0, 1, \ldots, N_E$. The value of i and j indicates the actions of selective caching and enhanced transmission functionalities, respectively. In the k-th time slot, action tuple (i, j) is further represented by $a(k) = (a_1(k), a_2(k))$, and the ingress node caches all the incoming packets whose layer number is greater than $a_1(k)$. To avoid a video rebuffering event (i.e., stalled video playback), the base-layer packets are not cached. When $a_1(k)$ is equal to 0, all the enhancement-layer packets arrived at the ingress node during the

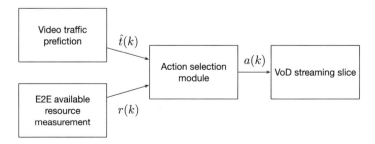

Fig. 4.14 The main elements of protocol framework

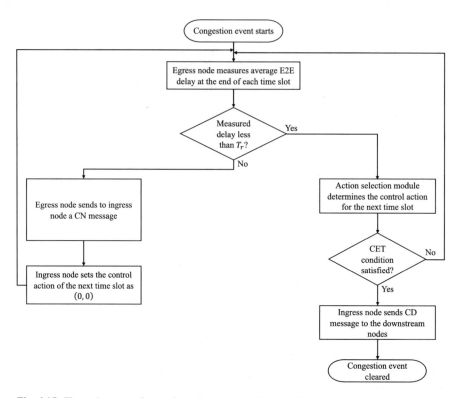

Fig. 4.15 The main protocol operation when a congestion event happens

k-th time slot are pushed into the caching buffer. If $a_1(k)$ equals N_e, no video packet needs to be cached in the k-th time slot. The value of $a_2(k)$ represents the number of packet chunks which should be transmitted by enhanced transmission functionality in the k-th time slot.

The main procedure of how the protocol is operated for the VoD streaming slice is illustrated in Fig. 4.15. At the end of the k-th time slot, the egress node measures average E2E delay $d_a(k)$. If $d_a(k)$ is greater than required delay bound T_r, the egress node enters the active mode and sends a `Congestion_Notification` (CN) message to the ingress node traversing backward the entire VoD streaming slice. An intermediate node in the VoD streaming slice changes to the active mode as soon as it receives a CN message. Once the action selection module at the ingress node receives the CN message, it sets the action of both selective caching and enhanced transmission functionalities as 0 for the $(k + 1)$-th time slot, i.e., $a(k + 1) = (0, 0)$. The purpose of caching all the enhancement-layer packets in the $(k + 1)$-th time slot is to reduce the queueing delay of the video packets to a maximal extent. The egress node measures $d_a(k + 1)$ which is included in a `Delay` message sent back to the ingress node. If $d_a(k + 1)$ is greater than the delay bound, the ingress node keeps caching all the enhancement-layer packets in the following time slots until

---

Algorithm 4: Protocol operation

1: **for** each time slot **do**
2:     Egress node measures the average E2E delay.
3:     **if** the measured delay is greater than T_r **then**
4:         Egress node sends CN message to the ingress node.
5:         Ingress node sets the action of selective caching functionality for the next time slot as 0.
6:         Ingress node sets the action of enhanced transmission functionality for the next time slot as 0.
7:     **else**
8:         Video traffic prediction module predicts the video traffic load for the next time slot.
9:         action selection module determines the action tuple of the next time slot.
10:    **end if**
11: **end for**

---

the average E2E delay is less than T_r. If the average E2E delay of the j-th time slot satisfies the delay requirement, the action selection module determines the action tuples of the following time slots, where the decision is made based on the predicted traffic load and the E2E available resources of the VoD streaming slice. The egress node measures and sends the feedback reward of executing the control action to the ingress node at the end of the time slot. The information is used to update the action-selection strategy. Since the egress node needs to update both the average E2E delay and the feedback reward at the end of each time slot during the active mode, the Delay message and Reward message can be encapsulated into one packet. While this packet passes through an intermediate node, the node attaches its current available resources information. The available resources measurement module at the ingress node uses this information from all the intermediate nodes in the slice to determine the E2E available resources. When the congestion event is over, the ingress node sends the cached packets to the corresponding video clients by enhanced transmission. Suppose the caching buffer at the ingress node becomes empty in the k-th time slot and $a_1(k)$ is N_e (i.e., no video packet is cached in the k-th time slot). The ingress node enters the deactivated mode and sends a CONTROL_DEACTIVATION (CD) message to the downstream nodes in the VoD streaming slice at the end of the k-th time slot. The condition of triggering the CD message is referred to as CET condition. A node changes to the deactivated mode when it receives a CD message. The egress node stops measuring the feedback reward and sending the Reward message until the next congestion event occurs in the network. The protocol operation for the VoD streaming slice in the active mode is summarized in Algorithm 4. The main functions of edge nodes in the VoD streaming slice are summarized in Table 4.2. The items followed by (all) are the functions required throughout the entire network operation, otherwise, the items are required only when the nodes are in the active mode.

Table 4.2 The functions of the edge nodes in the VoD streaming slice

Node type	Functions
Ingress node	– Control action selection and execution – E2E available resources measurement – Video traffic prediction – Sending CD messages
Egress node	– Average E2E delay measurement (all) – Feedback reward measurement – Sending CN messages – Sending Delay and Reward messages

Next, we describe the mechanism of managing the caching buffer at the ingress node. The caching buffer is operated in the FIFO manner. To better use the caching resources, the caching buffer drops the packets from the video segments which have been played out by the clients. The video clients periodically report the buffer information to the SDN/NFV-enabled controller of the core network, containing the segment number of video segment being played out [40]. Then, the controller forwards this information to the ingress node of the VoD streaming slice. When the caching buffer receives the message of buffer information, it removes the packets of the same client whose segment number is less than or equal to the segment number indicated in the message.

4.3.6 Video Traffic Prediction

The video traffic prediction module in Fig. 4.14 is used to forecast the video traffic load in each time slot. Since the congestion control action selection is conducted at different encoded video layers, the traffic load in each time slot is predicted at different SVC layers. The maximum number of enhancement-layers, N_e, of all the video files stored at the video server is assumed to be identical. Thus, the output dimension from the video traffic prediction module is $N_e + 1$. The prediction result for the k-th time slot is expressed as

$$\hat{\mathbf{t}}(k) = \left[\hat{t}_0(k), \hat{t}_1(k), \hat{t}_2(k), \dots, \hat{t}_{N_e}(k) \right] \qquad (4.29)$$

where $\hat{t}_i(k)$ represents the predicted number of packet arrivals of layer i in the k-th time slot. The predicted traffic load of base-layer packets is denoted by $\hat{t}_0(k)$. Note that we only need to have one traffic prediction module at the ingress node which are fed with the information of each SVC layer to obtain layer-level traffic prediction results. The ARIMA model is used for video traffic prediction, which takes the traffic load of the past time slots as input and predicts the amount of packet arrivals in the next time slot [41, 42].

4.3.6.1   Model Parameters

The ARIMA model is specified by three parameters d, p and q, where d is the degree of differencing (i.e., the number of differencing to eliminate the trend of a non-stationary time series), p is the order of the autoregressive model, and q is the order of the moving-average model. The parameters can be determined by analyzing the historical traffic load patterns. Denote by $h_i(k)$ the observed traffic load of layer i (i.e. the number of video packets of layer i arrived at the ingress node) during the k-th time slot. The time series of the historical traffic load is represented by $\{h_i(k)\}$. Let $h_i(T)$ denote a vector of traffic loads observed in T time slots, given by

$$h_i(T) = \Big[h_i(1), h_i(2), \ldots, h_i(T) \Big]. \tag{4.30}$$

Let $\nabla^c h_i(T)$ denote the c-th-order difference of $h_i(T)$. The value of c is determined by conducting the augmented Dickey-Fuller (ADF) test for $\nabla^c h_i(T)$ ($c = 0, 1, \ldots$) [41, 43]. If the $p\text{-}value$[3] for $\nabla^c h_i(T)$ is less than a pre-determined threshold (e.g., 0.05), the corresponding time series, $\{\nabla^c h_i(k)\}$, is stationary. Then, parameter d is set as c, otherwise, more differencing is required to transform $\{\nabla^c h_i(k)\}$ to a stationary time series. Given d, the selection of parameters p and q is based on the minimization of the corrected Akaike information criterion statistic [41]. Time series $\{\nabla^d h_i(k)\}$ being stationary indicates that its mean is constant. Denote by μ_i the sample mean of $\nabla^d h_i(T)$.

4.3.6.2   Traffic Prediction via ARIMA Model

Given d, p and q, the ARIMA model predicts the traffic load at the beginning of each time slot during the network operation. The vector of the observed traffic loads for the first $d + k$ time slots during the network operation is expressed as

$$t_i(d + k) = \Big[t_i(1), \ldots, t_i(d), t_i(d + 1), \ldots, t_i(d + k) \Big]. \tag{4.31}$$

Let $\nabla^d t_i(d + k)$ denote the d-th-order difference of $t_i(d + k)$, which is represented as

$$\nabla^d t_i(d + k) = \Big[\nabla^d t_i(d + 1), \nabla^d t_i(d + 2), \ldots, \nabla^d t_i(d + k) \Big]. \tag{4.32}$$

---

[3] The p-value is used in statistical test for determining whether to reject the null hypothesis, which is different with the above-mentioned parameter p.

From [41], the predicted traffic load of layer i in the $(d+k+1)$-th time slot is given by

$$\hat{t}_i(d+k+1) = \widehat{\nabla^d t_i}(d+k+1)$$

$$- \sum_{j=1}^{d} \binom{d}{j}(-1)^j t_i(d+k+1-j)$$

where $\widehat{\nabla^d t_i}(d+k+1)$ is the prediction of $\nabla^d t_i(d+k+1)$ given $\nabla^d t_i(d+k)$. Now, the traffic prediction problem becomes how to determine $\widehat{\nabla^d t_i}(d+k+1)$. We define $y_i(k)$ as

$$y_i(k) = \left[y_i(1), y_i(2), \ldots, y_i(k) \right] \tag{4.33}$$

where $y_i(j)$ $(j = 1, 2, \ldots, k)$ is equal to $\nabla^d t_i(d+j) - \mu_i$. Let $\hat{y}_i(k+1)$ denote the prediction of $y_i(k+1)$. Since μ_i is estimated before the network operation begins based on the historical traffic load patterns, the traffic prediction problem is finally converted to determining $\hat{y}_i(k+1)$ given $y_i(k)$. The recursive equation of finding the value of $\hat{y}_i(k+1)$ is given by

$$\hat{y}_i(k+1) = \begin{cases} \sum_{j=1}^{k} \theta_{k,j}[y_i(k+1-j) \\ \quad - \hat{y}_i(k+1-j)], \ 1 \le k < v \\ \sum_{j=1}^{q} \theta_{k,j}[y_i(k+1-j) - \hat{y}_i(k+1-j)] \\ \quad + \alpha_1 y_i(k) + \cdots + \alpha_p y_i(k+1-p), \ k \ge v \end{cases} \tag{4.34}$$

where v is the maximum of p and q [41, 42]. Note that $\hat{y}_i(1)$ equals 0. The coefficients in (4.34) (i.e., $\alpha_1, \ldots, \alpha_p, \theta_{k,j}$) can be calculated recursively as in [41].

4.3.7 MAB Learning Based Action Selection

The deployment of selective caching and enhanced transmission functionalities at the ingress node from a VoD streaming slice is to deliver more video packets without leading to network congestion. Therefore, we define the reward of executing action $a(k)$ in the k-th time slot as

$$R_{a(k)}(k) = \frac{g(k)}{r(k)T_s} \tag{4.35}$$

where $g(k)$ is the number of video packets sent from the VoD streaming slice in the k-th time slot within the required delay bound T_r. Through triggering different actions in each time slot, the ingress node intends to maximize the expected overall reward over time.

The per-slot reward of executing an action under different network conditions may be different. Caching video packets during a congestion event can reduce the packet E2E delay to increase the reward, whereas the reward is decreased if the ingress node activates the selective caching functionality when there is no congestion event. Therefore, the video traffic load and E2E available resources of the VoD streaming slice should be taken into consideration when the action selection module selects the control actions at each time slot. We formulate this action-selection problem as an MAB problem, where the predicted video traffic load and E2E available resources are treated as the context information. The MAB problem which uses context information for decision making is also referred to as contextual bandit problem [31], where the selected arms are control actions at each time slot.

Let x_k denote the context information at the k-th time slot, given by

$$x_k = \left[\hat{t}_0(k), \hat{t}_1(k), \hat{t}_2(k), \ldots, \hat{t}_{N_e}(k), r(k) \right]. \tag{4.36}$$

The LinUCB algorithm is employed to solve the MAB problem with context information [31]. For the k-th time slot, the expected reward of an action $a \in \mathcal{A}$ is expressed as

$$E\left[R_a(k) | x_k \right] = x_k^T \theta_a^* \tag{4.37}$$

where θ_a^* is an unknown coefficient vector. Assume m contexts of action a have been observed before the k-th time slot and the corresponding feedback rewards are recorded by response vector $R_a \in \mathbb{R}^m$. The matrix of the m observed contexts for action a is denote by $D_a \in \mathbb{R}^{m \times z}$, where z is the dimension of the context (i.e., $N_e + 2$). The estimate of the coefficient vector, θ_a^*, is given by

$$\hat{\theta}_a = \left(D_a^T D_a + I_z \right)^{-1} D_a^T R_a \tag{4.38}$$

where I_z is the $z \times z$ identity matrix. It is shown in [31] that, for any $\delta > 0$, the following inequality holds with a probability of at least $1 - \delta$,

$$\left| x_k^T \hat{\theta}_a - E\left[R_a(k) | x_k \right] \right| \leq \xi \sqrt{x_k^T A_a x_k} \tag{4.39}$$

where

$$A_a = D_a^T D_a + I_z, \quad \xi = 1 + \sqrt{\frac{\ln(2/\delta)}{2}}. \tag{4.40}$$

Algorithm 5: Action-selection algorithm

1: Initialize $\xi \in \mathbb{R}_+$ and $d_a(0) = 0$.
2: **for** $k = 1, 2, \ldots$ **do**
3:　　**if** $d_a(k-1) > T_r$ **then**
4:　　　　Set the action tuple of the k-th time slot as $(0, 0)$.
5:　　**else**
6:　　　　Obtain the context information: $x_k \in \mathbb{R}^z$.
7:　　　　**for** every $a \in \mathcal{A}$ **do**
8:　　　　　　**if** a is new **then**
9:　　　　　　　　$A_a \leftarrow I_z$
10:　　　　　　　$b_a \leftarrow 0_{z \times 1}$
11:　　　　　　**end if**
12:　　　　　$\hat{\theta}_a \leftarrow A_a^{-1} b_a$
13:　　　　　$\hat{R}_a(k) \leftarrow x_k^T \hat{\theta}_a + \xi \sqrt{x_k^T A_a^{-1} x_k}$
14:　　　　**end for**
15:　　　　Set the action tuple of the k-th time slot $a(k) = \arg\max_{a \in \mathcal{A}} \hat{R}_a(k)$.
16:　　　　Observe the actual reward $R_{a(k)}(k)$ at the end of k-th time slot.
17:　　　　$A_a \leftarrow A_a + x_k x_k^T$
18:　　　　$b_a \leftarrow b_a + R_{a(k)}(k) x_k$
19:　　**end if**
20: **end for**

At the beginning of the k-th time slot, the action selection module selects the action which maximizes $\hat{R}_a(k)$ as

$$\hat{R}_a(k) = x_k^T \hat{\theta}_a + \xi \sqrt{x_k^T A_a^{-1} x_k}. \qquad (4.41)$$

Recall that the selected control action in the k-th time slot is denoted by $a(k)$. The actual reward, $R_{a(k)}(k)$, of taking action $a(k)$ in the k-th time slot is observed at the end of the slot. Then, the tuple, $(x_k, a(k), R_{a(k)}(k))$, is fed back to the action selection module to update the parameters of the LinUCB algorithm. The detailed action-selection algorithm when the VoD streaming slice is in the active mode is presented in Algorithm 5.

4.3.8   Performance Evaluation

In this section, performance evaluation of the proposed protocol is conducted, where four QoS performance metrics are considered in the evaluation, i.e., average E2E delay, throughput, goodput ratio, and resource utilization.

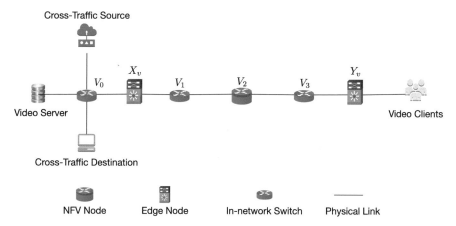

Fig. 4.16 Network topology for simulation

Table 4.3 Packet arrival rate of the cross-traffic at V_0

Time interval	[1, 40]	[41, 80]	[81, 120]
	1100	1400	1700
Packet arrival rate	packet/s	packet/s	packet/s

4.3.8.1 Simulation Settings

The network topology for simulation is shown in Fig. 4.16. There are five video clients downloading video files from the same video server [39]. The duration of a video segment, Δ_s, of all video files is 2 s [44]. Every segment is encoded into one base-layer chunk and four enhancement-layer chunks [45]. Each video chunk is delivered by 200 video packets, and the packet size is constant, equals to 1400 bytes [39]. The aggregated video traffic flow passes through switch V_0 to ingress node X_v. Nodes X_v and Y_v are the ingress node and the egress node for the VoD streaming slice, respectively. The VoD streaming slice between the edge nodes has a linear topology which contains two in-network switches (i.e., V_1 and V_3) and one NFV node (i.e., V_2). As discussed in Sect. 4.3.1, the edge nodes are in-network servers which have much more resources than NFV nodes and in-network switches. The capacity of node V_l ($l = 0, 1, 2, 3$) is $C_l = 4500$ packet/s [46]. The video traffic flow and the cross-traffic flow share the transmission resources at V_0. During the network operation, we change the packet arrival rate of the cross-traffic at V_0 according to the settings in Table 4.3 to evaluate the performance of the proposed protocol. The loss-based congestion control algorithm based on the additive-increase multiplicative-decrease (AIMD) mechanism is implemented at the video server to control the source sending rate [47]. The propagation delay of the links outside (between) the edge nodes is set as 5 ms (2.5 ms). The E2E delay bound T_r is set to 40 ms. Parameter ξ in (4.40) is 1.5 [31]. The total simulated slice time is 120 s and the length of every time slot is 1 s. No cached packet is dropped during the network operation.

Table 4.4 ADF test results when $d = 0$

ADF statistic	p-value	Critical value (1%)	Critical value (5%)	Critical value (10%)
-0.913	0.784	-3.489	-2.887	-2.580

Table 4.5 ADF test results when $d = 1$

ADF statistic	p-value	Critical value (1%)	Critical value (5%)	Critical value (10%)
-9.627	1.647×10^{-16}	-3.489	-2.887	-2.580

We first determine parameters p, q and d of the ARIMA model for video traffic load prediction. The ingress node collects the video traffic loads of 120 time slots for data analysis. The augmented Dickey-Fuller (ADF) test is used to check the video traffic stationarity over time [41, 43]. The test results when $d = 0$ are given in Table 4.4. Since the p-value of the traffic load series is greater than a pre-determined threshold set as 0.05 [48], the video traffic load series when $d = 0$ is not stationary. We conduct the ADF test when $d = 1$, and the test results given in Table 4.5 indicate that the p-value is much less than the threshold, 0.05. In addition, the ADF statistic is less than all the critical values, indicating that the time series is stationary with a 99% confidence level [48]. Thus, parameter d is set as 1 in the simulation. Then, we select parameters p and q by evaluating the AICC statistic. Based on the observed traffic loads, the AICC statistic is minimized when $p = 2$ and $q = 1$. Therefore, the ARIMA model with $d = 1$, $p = 2$, and $q = 1$ is used for the video traffic prediction.

4.3.8.2   Numerical Results

The average E2E delay, goodput ratio, throughput, and resource utilization for the VoD streaming systems are compared between the proposed protocols and benchmark schemes. For brevity, we denote the VoD streaming system with (without) the proposed edge-operated transmission control protocol by VS-W (VS-WO) system. Figures 4.17 and 4.18 show the average E2E delay and goodput ratio performance, respectively. The results are obtained from ten repeated simulations. The throughput and resource utilization of each time slot in one simulation are presented in Figs. 4.19 and 4.20 respectively.

Average E2E Delay We first examine the average E2E delay of VS-W and VS-WO when a congestion event occurs in the VoD streaming slice. The network congestion is generated by reducing the available resources of V_2 from 4500 packet/s to 2500 packet/s. The resource of V_2 changes with time as specified in Table 4.6. Two congestion durations, 20 s and 40 s, are considered in the simulation. The cumulative distribution function (CDF) of the average E2E delay is measured and shown in Fig. 4.17. It can be seen that the CDFs of VS-W with different congestion intervals are close to each other. Also, the average E2E delay of the VS-W system measured in each time slot is less than the required delay bound, since the selective caching

Fig. 4.17 The evaluation of
average E2E delay

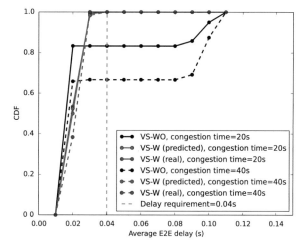

Fig. 4.18 The evaluation of
goodput ratio

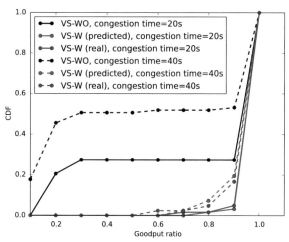

functionality is activated after the congestion happens. By caching a certain number
of enhancement-layer packets, the queue length at V_2 is under control. Figure 4.17
also shows the results of VS-W system whose action selection module is fed with the
predicted and real traffic load, respectively, for each time slot. The results obtained
based on predicted and real video traffic load information are similar, thanks to the
effectiveness of the traffic prediction algorithm. For VS-WO, the average E2E delay
in around 17% time slots exceeds the required delay bound due to a 20 s congestion
event. In around 32% time slots, the delay requirement is not satisfied when a 40 s
congestion event occurs. Thus, the performance gap between VS-W and VS-WO
systems becomes larger if the network congestion lasts longer. It is observed that
the CDF of VS-WO is greater than that of VS-W when the average E2E delay is
0.02 s. For VS-WO, the queueing delay is negligible after the network congestion.
Hence, the average E2E delay within these time slots is in the range between 0.01 s

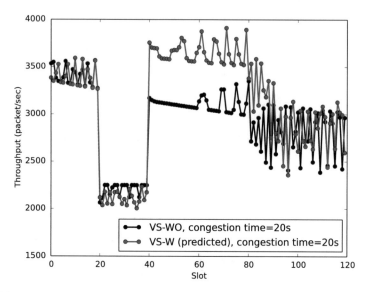

Fig. 4.19 Throughput with respect to the number of slots

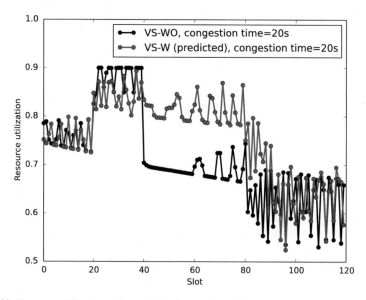

Fig. 4.20 Resource utilization with respect to the number of slots

and 0.02 s. However, the enhanced transmission functionality is activated in the VS-W system after network congestion. As a result, the average E2E delay within the corresponding time slots increases to some extent, but does not exceed the required delay bound.

Table 4.6 The available resources of V_2 over time

Time interval	[1, 20]	[21, 40]	[41, 60]	[61, 120]
Congestion duration 20 s	4500 packet/s	2500 packet/s	4500 packet/s	4500 packet/s
Congestion duration 40 s	4500 packet/s	2500 packet/s	2500 packet/s	4500 packet/s

Goodput Ratio In evaluating the goodput ratio, the available resources of V_2 is set as in Table 4.6. It can be seen from Fig. 4.18 that VS-W outperforms VS-WO in different congestion intervals. Furthermore, the goodput ratio of VS-W is not sensitive to the congestion durations. The performance gap between VS-W and VS-WO systems increases with the congestion time. We also compare the performance of VS-W systems with the predicted traffic load and with the real traffic load, which are close to each other as expected.

Throughput and Resource Utilization To validate the effectiveness of the proposed enhanced transmission functionality, we compare the throughput of VS-W and VS-WO systems at each time slot during the network operation, as shown in Fig. 4.19. The action selection module of VS-W system utilizes the predicted traffic load information in action selection. The congestion event occurs at V_2 from 20 s to 40 s. Before the network congestion, the throughputs of the VS-W and VS-WO systems are close to each other, since they depend only on the video traffic load. During the network congestion, the throughput of two VoD streaming systems is also at the same level. The network congestion is mitigated after the 40-th time slot and the ingress node of VS-W starts to send cached video packets to the corresponding video clients by enhanced transmission functionality. Therefore, the throughput of VS-W is higher than that of VS-WO from the 41-th time slot. All the cached video packets are transmitted before the 91-th time slot. As expected, the throughput of VS-W returns to the same level of VS-WO from the 91-th time slot to the end of the simulation. Figure 4.20 shows the resource utilization for the two VoD streaming systems. With the enhanced transmission functionality, the resource utilization of VS-W is higher than that of VS-WO from the 41-th time slot to the 90-th time slot.

4.4 Summary

In this chapter, we have presented SDN/NFV-enabled transmission protocols customized for time-critical and VoD streaming services, respectively, in a 5G core network. For supporting the time-critical service, we propose the SDATP with in-network intelligence enabled, such as in-path caching-based retransmission and in-network caching-based congestion control, which achieves early packet loss recovery and congestion mitigation. The placement of caching functionalities and the packet caching probabilities are jointly optimized to improve the SDATP performance; For supporting the video streaming service, we have further proposed a customized transmission protocol with in-network congestion control and throughput enhancement functionalities to realize fast reaction to network dynamics. To balance the tradeoff between congestion control and QoS provisioning, an MAB problem is formulated to maximize the overall network performance over time by triggering proper control actions under different network conditions, with the consideration of predicted video traffic load information and E2E available resources. Performance evaluation is conducted to demonstrate the advantages of the proposed customized protocols in terms of both congestion alleviation and QoS provisioning.

References

1. I. Parvez, A. Rahmati, I. Guvenc, A.I. Sarwat, H. Dai, A survey on low latency towards 5G: RAN, core network and caching solutions. IEEE Commun. Surv. Tuts. **20**(4), 3098–3130 (2018)
2. V.G. Nguyen, A. Brunstrom, K.J. Grinnemo, J. Taheri, SDN/NFV-based mobile packet core network architectures: a survey. IEEE Commun. Surv. Tuts. **19**(3), 1567–1602 (2017)
3. J. Postel, *Transmission Control Protocol,* RFC 793, IETF, Sept. 1981.
4. D. Borman, B. Branden, V. Jacobson, R. Scheffenegger, *TCP Extensions for High Performance,* RFC 7323, IETF, Sept. 2014.
5. E. Miravalls-Sierra, D. Muelas, J. Ramos, J.E.L. de Vergara, D. Morató, J. Aracil, Online detection of pathological TCP flows with retransmissions in high-speed networks. Comput. Commun. **127**, 95–104 (2018)
6. S. Kadry, A.E. Al-Issa, Modeling and simulation of out-of-order impact in TCP protocol. J. Adv. Comput. Netw. **3**(3), 220–224 (2015)
7. J. Zhou, Z. Li, Q. Wu, P. Steenkiste, S. Uhlig, J. Li, G. Xie, TCP stalls at the server side: measurement and mitigation. IEEE/ACM Trans. Netw. **27**(1), 272–287 (2019)
8. Y. Edalat, J.-S. Ahn, K. Obraczka, Smart experts for network state estimation. IEEE Trans. Netw. Service Manag. **13**(3), 622–635 (2016)
9. P. Yang, J. Shao, W. Luo, L. Xu, J. Deogun, Y. Lu, TCP congestion avoidance algorithm identification. IEEE/ACM Trans. Netw. **22**(4), 1311–1324 (2013)
10. S. Ferlin, Ö. Alay, T. Dreibholz, D.A. Hayes, M. Welzl, Revisiting congestion control for multipath TCP with shared bottleneck detection, in *Proc. IEEE INFOCOM*, 2016, pp. 1–9
11. S.P. Tinta, A.E. Mohr, J.L. Wong, Characterizing end-to-end packet reordering with UDP traffic, in *Proc. ISCC*, 2009, pp. 321–324
12. K. Bao, J.D. Matyjas, F. Hu, S. Kumar, Intelligent software-defined mesh networks with link-failure adaptive traffic balancing. IEEE Trans. Cogn. Commun. Netw. **4**(2), 266–276 (2018)
13. H. Soni, W. Dabbous, T. Turletti, H. Asaeda, NFV-based scalable guaranteed-bandwidth multicast service for software defined ISP networks. IEEE Trans. Netw. Service Manag. **14**(4), 1157–1170 (2017)
14. N. Zhang, P. Yang, S. Zhang, D. Chen, W. Zhuang, B. Liang, X. Shen, Software defined networking enabled wireless network virtualization: challenges and solutions. IEEE Netw. **31**(5), 42–49 (2017)
15. Q. Ye, W. Zhuang, X. Li, J. Rao, End-to-end delay modeling for embedded VNF chains in 5G core networks. IEEE Internet Things J. **6**(1), 692–704 (2019)
16. M.R. Sama, L.M. Contreras, J. Kaippallimalil, I. Akiyoshi, H. Qian, H. Ni, Software-defined control of the virtualized mobile packet core. IEEE Commun. Mag. **53**(2), 107–115 (2015)
17. K. Qu, W. Zhuang, Q. Ye, X. Shen, X. Li, J. Rao, Delay-aware flow migration for embedded services in 5G core networks, in *Proc. IEEE ICC*, Shanghai, China, 2019, pp. 1–6

18. O. Alhussein, P.T. Do, Q. Ye, J. Li, W. Shi, W. Zhuang, X. Shen, X. Li, J. Rao, A virtual network customization framework for multicast services in NFV-enabled core networks. IEEE J. Sel. Areas Commun. **38**(6), 1025–1039 (2020)

19. K. Qu, W. Zhuang, Q. Ye, X. Shen, X. Li, J. Rao, Dynamic flow migration for embedded services in SDN/NFV-enabled 5G core networks. IEEE Trans. Commun. **68**(4), 2394–2408 (2020)

20. E. Soltanmohammadi, K. Ghavami, M. Naraghi-Pour, A survey of traffic issues in machine-to-machine communications over LTE. IEEE Internet Things J. **3**(6), 865–884 (2016)

21. Y. Liu, C. Yuen, X. Cao, N.U. Hassan, J. Chen, Design of a scalable hybrid MAC protocol for heterogeneous M2M networks. IEEE Internet Things J. **1**(1), 99–111 (2014)

22. K.P. Choi, On the medians of gamma distributions and an equation of Ramanujan. Proc. Am. Math. Soc. **121**(1), 245–251 (1994)

23. J. Chen, S. Yan, Q. Ye, W. Quan, P.T. Do, W. Zhuang, X. Shen, X. Li, J. Rao, An SDN-based transmission protocol with in-path packet caching and retransmission, in *Proc. IEEE ICC*, 2019, pp. 1–6

24. B. Davie, J. Gross, A stateless transport tunneling protocol for network virtualization (STT) (2016). *Internet Engineering Task Force*. https://tools.ietf.org/html/draft-davie-stt-08

25. R. Hemmecke, M. Köppe, J. Lee, R. Weismantel, Nonlinear integer programming, in *50 Years of Integer Programming 1958–2008* (Springer, Berlin, 2009), pp. 561–618

26. B. Lantz, B. Heller, N. McKeown, A network in a laptop: rapid prototyping for software-defined networks, in *Proceedings of the 9th ACM SIGCOMM Workshop on Hot Topics in Networks* (2010), pp. 1–6

27. Open vswitch. http://www.openvswitch.org/

28. Ryu SDN framework. http://osrg.github.io/ryu/

29. B. Pfaff, B. Lantz, B. Heller, Openflow switch specification, version 1.3. 0, in *Open Networking Foundation* (2012)

30. H.T. Nguyen, J. Mary, P. Preux, Cold-start problems in recommendation systems via contextual-bandit algorithms (2014). arXiv preprint arXiv:1405.7544

31. L. Li, W. Chu, J. Langford, R.E. Schapire, A contextual-bandit approach to personalized news article recommendation, in *Proc. ACM WWW* (Raleigh, 2010), pp. 661–670

32. M. Moradi, W. Wu, L.E. Li, Z.M. Mao, SoftMoW: recursive and reconfigurable cellular WAN architecture, in *Proc. ACM CONEXT* (Sydney, 2014), pp. 377–390

33. J. Chen, Q. Ye, W. Quan, S. Yan, P.T. Do, W. Zhuang, X. Shen, X. Li, J. Rao, SDATP: an SDN-based adaptive transmission protocol for time-critical services. IEEE Netw. **34**(3), 154–162 (2020)

34. P. Megyesi, A. Botta, G. Aceto, A. Pescape, S. Molnar, Available bandwidth measurement in software defined networks, in *Proc. ACM SAC* (Pisa, 2016), pp. 651–657

35. Z.-L. Zhang, Y. Wang, D.H.-C. Du, D. Su, Video staging: a proxy-server-based approach to end-to-end video delivery over widearea networks. IEEE/ACM Trans. Netw. **8**(4), 429–442 (2000)

36. H. Schwarz, D. Marpe, T. Wiegand, Overview of the scalable video coding extension of the H. 264/AVC standard. IEEE Trans. Circuits Syst. Video Technol. **17**(9), 1103–1120 (2007)

37. T. Stockhammer, Dynamic adaptive streaming over HTTP: standards and design principles, in *Proc. ACM MMSYS* (San Jose, 2011), pp. 133–144

38. T.-Y. Huang, R. Johari, N. McKeown, M. Trunnell, M. Watson, A buffer-based approach to rate adaptation: evidence from a large video streaming service, in *Proc. ACM SIGCOMM* (Chicago, 2014), pp. 187–198

39. S. Yan, P. Yang, Q. Ye, W. Zhuang, X. Shen, X. Li, J. Rao, Transmission protocol customization for network slicing: a case study of video streaming. IEEE Veh. Technol. Mag. **14**(4), 20–28 (2019)

40. C.W. Chen, P. Chatzimisios, T. Dagiuklas, L. Atzori, *Multimedia Quality of Experience (QoE): Current Status and Future Requirements* (Wiley, London, 2015)

41. P.J. Brockwell, R.A. Davis, *Introduction to Time Series and Forecasting* (Springer, Berlin, 2016)

42. A. Azzouni, G. Pujolle, NeuTM: a neural network-based framework for traffic matrix prediction in SDN, in *Proc. IEEE/IFIP NOMS* (Taipei, 2018), pp. 1–5
43. A. Pal, P. Prakash, *Practical Time Series Analysis: Master Time Series Data Processing, Visualization, and Modeling Using Python* (Packt Publishing, 2017)
44. S. Garcia, J. Cabrera, N. Garcia, Quality-control algorithm for adaptive streaming services over wireless channels. IEEE J. Sel. Topics Signal Process. **9**(1), 50–59 (2014)
45. M. Rahmati, D. Pompili, UW-SVC: scalable video coding transmission for in-network underwater imagery analysis, in *Proc. IEEE MASS*, Monterey, 2019, pp. 380–388
46. O. Alhussein, W. Zhuang, Robust online composition, routing and NF placement for NFV-enabled services. IEEE J. Sel. Areas Commun. **38**(6), 1089–1101 (2020)
47. B. Sikdar, S. Kalyanaraman, K.S. Vastola, Analytic models for the latency and steady-state throughput of TCP Tahoe, Reno, and SACK. IEEE/ACM Trans. Netw. **11**(6), 959–971 (2003)
48. J.M. Weiming, *Mastering Python for Finance* (Packt Publishing, 2015)

Chapter 5
Adaptive Medium Access Control Protocols for IoT-Enabled Mobile Networks

5.1 Load-Adaptive MAC with Homogeneous Traffic

IoT-enabled mobile networks have been envisioned as an important revolution for future generation wireless networks, which facilitate a growing number of smart heterogeneous objects (e.g., smart sensors, home appliances, and monitoring devices) being connected via suitable wireless technologies for ubiquitous information dissemination and seamless communication interaction [1]. A mobile ad hoc network (MANET) is one typical example of IoT-enabled mobile networks, allowing a group of smart mobile devices (e.g., smartphones, and smart sensor nodes) interconnected in a peer-to-peer manner and communicating without relying on any network infrastructure. MANETs are widely deployed for pervasive IoT-oriented applications, for example, smart home/office networking [1], emergency communications and crisis response in disaster areas [2–4], and tactical networks for the purpose of command interactions [5]. Device-to-Device (D2D) communication networks have been recently popularized as one of the typical realizations for IoT-enabled MANETs, which rely on the ad hoc networking of spatially-distributed smart devices for information relaying and sharing in certain areas where conventional communication infrastructures are not accessible [3, 4, 6].

For a typical IoT-enabled MANET with power-rechargeable mobile nodes [7] (e,g., smartphones, and smart sensors) generating a high volume of heterogeneous traffic, supporting heterogeneous services with differentiated QoS guarantee becomes an important but challenging task. The network is expected to not only provide as high as possible throughput for best-effort data traffic, but also ensure a bounded packet loss rate for delay-sensitive voice communications or even multimedia streaming. Therefore, a QoS-aware MAC protocol is required to coordinate the channel access behaviors from end devices/users of heterogeneous

© The Author(s), under exclusive license to Springer Nature Switzerland AG 2021
Q. Ye, W. Zhuang, *Intelligent Resource Management for Network Slicing in 5G and Beyond*, Wireless Networks, https://doi.org/10.1007/978-3-030-88666-0_5

service types to satisfy their differentiated QoS requirements [8]. However, the characteristics of MANETs pose technical challenges in the QoS-aware MAC design: (1) Since MANETs do not depend on any central control, distributed MAC is required to coordinate the transmissions of neighboring nodes based on their local information exchanges; (2) Nodes are mobile, making the heterogeneous network traffic load change with time. The traffic load variations can lead to QoS performance degradation. Thus, MAC is expected to be context-aware, which adapts to the changing network traffic load to achieve consistently satisfactory service performance.

In this section, we first present an adaptive MAC solution for a fully-connected MANET supporting homogeneous best-effort data traffic, which is then extended to a scenario with heterogeneous traffic in Sect. 5.2. Specifically, based on the detection of current network load condition, nodes can make a switching decision between IEEE 802.11 distributed coordination function (DCF) and dynamic time division multiple access (D-TDMA), when the network traffic load reaches a threshold, referred to as MAC switching point. The adaptive MAC solution determines the MAC switching point to maximize the network performance. Approximate and closed-form performance analytical models for both MAC protocols are established, which facilitate the computation of MAC switching point in a tractable way.

5.1.1 Network Model

Consider a fully-connected MANET [9–11] with a single and error-free channel [12, 13]. There is no central controller in the network, and nodes coordinate their transmissions in a distributed way. The destination node for each source node is randomly selected from the rest nodes. Each mobile node generates best-effort data traffic. The data traffic arrivals at each node are modeled as a Poisson process with an average arrival rate λ packet/s [14–16]. Packet loss among any pair of source-destination (S-D) nodes results from packet transmission collisions. The total number of nodes in the network is denoted by N, which changes slowly with time (with respect to a packet transmission time), due to user/device mobility.

5.1.2 Adaptive MAC Framework

Consider two candidate MAC protocols maintained at each node in the adaptive MAC framework [17], in which a separate mediating MAC entity working on top of the MAC candidates can reconfigure the current MAC layer by switching from one MAC protocol to the other, based on the current network condition (e.g., interference level, and the total number of nodes). This adaptation of MAC to

Fig. 5.1   The frame structure of D-TDMA

the networking environment has a potential to improve the network performance. The contention-based IEEE 802.11 DCF is considered as one candidate MAC protocol, which is a standardized and widely adopted MAC scheme based on the carrier sense multiple access with collision avoidance (CSMA/CA). It has a better channel utilization than slotted-Aloha [18], and has high performance at a low contention level. Since we consider a fully-connected MANET scenario where no hidden terminal problem exists, the basic access mechanism in IEEE 802.11 DCF is considered.

The channelization-based D-TDMA scheme [19] is chosen as the other MAC candidate, which is originally used in cellular networks. Time is partitioned to frames of constant duration. Each D-TDMA frame consists of a control period and data packet transmission period. The control period has a number of constant-duration minislots, and data transmission period is composed of a number of equal-length data slots. The duration of each data slot is the time used to transmit one data packet. The number of minislots indicates the maximum number of users the network can support, and the number of data slots equals current total number of nodes, N, in the network. The D-TDMA frame structure is shown in Fig. 5.1. In order to fit the distributed MANET scenario, the minislots in the control period of each D-TDMA frame is used for local information exchange and distributed data slot acquisition of each node. The D-TDMA can support a varying number of nodes in the network and achieve high channel utilization in a high data traffic load.

5.1.3   Closed-Form Performance Models for IEEE 802.11 DCF and D-TDMA

In this subsection, a unified performance analysis framework is established for both candidate MAC protocols. We present approximate and closed-form expressions for the relation between performance metrics (i.e., network throughput and packet delay) and the total number of nodes in the network. Both traffic saturation and non-saturation cases are considered. All the time durations of IEEE 802.11 DCF are normalized to the unit of a back-off time slot in the IEEE 802.11b standard.

5.1.3.1 Closed-Form Performance Models for IEEE 802.11 DCF

In a homogeneous traffic case, because of the throughput fairness property of IEEE 802.11 DCF [20–22], the network throughput[1] definition can be made over one renewal cycle[2] of the transmission process. It is defined as the ratio of average payload transmission duration during one renewal cycle over the average length of the cycle [16], given by

$$S = \frac{N T_{pl}}{N \left(T_s + \frac{\overline{T_c}}{2} \right) + \overline{CW_2} + (1 - \rho_r)[1 - (N - 1)\rho_r]\left(\frac{1}{\lambda} - D_T \right)}. \tag{5.1}$$

In (5.1), T_{pl} is the duration of each packet payload; T_s is the successful transmission time of one packet; $\overline{T_c} = \frac{p}{1-p} T_c$ is the average collision time encountered by each packet before it is successfully transmitted [12], assuming a large retransmission limit, T_c is the collision time that each packet experiences when a collision occurs, p is the packet collision probability conditioned on that the node attempts a transmission, which is assumed to be constant and independent of the number of retransmissions; $\overline{CW_2} = \frac{W_0}{2} + p\frac{W_1}{2} + p^2\frac{W_2}{2} + \ldots + p^{M_b}\frac{W_{M_b}}{2} + p^{M_b+1}\frac{W_{M_b}}{2} + \ldots + p^{M_L}\frac{W_{M_b}}{2}$ denotes the average back-off contention window time spent by the tagged node i during the cycle, where $W_j = 2^j CW (j = 0, 1, \ldots, M_b)$ is the back-off contention window size in the j-th back-off stage (CW is the minimum contention window size), M_b is the maximum back-off stage, M_L is the retransmission limit; $\rho_r = \frac{\lambda}{\mu_s} = \frac{\lambda}{N\mu_d}$ is the probability with which an incoming packet sees a non-empty queue [16], where μ_s denotes the average service rate of the IEEE 802.11 DCF, μ_d is the average packet service rate seen by an individual node; and D_T is the average packet delay, defined as the duration from the instant that a packet arrives at the transmission queue to the instant that the packet is successfully transmitted, averaged over all transmitted packets of each node.

Performance Analysis in a Traffic Saturation Case In a traffic saturation case, (5.1) can be simplified to represent the saturation throughput S_1, given by

$$S_1 = \frac{N T_{pl}}{N \left(T_s + \frac{\overline{T_c}}{2} \right) + \overline{CW_2}} \tag{5.2}$$

which is a function of the number of nodes, N, and conditional collision probability p [23]. The collision probability p is correlated with N [24, 25],

$$p = 1 - (1 - \tau)^{N-1} \tag{5.3}$$

---

[1] The throughput in this section is normalized by the channel rate.

[2] The transmission attempt process of each node can be regarded as a regenerative process with the renewal cycle being the time between two successfully transmitted packets of the node [16].

where τ is the packet transmission probability of each node in any back-off time slot given a nonempty queue, and can be also expressed as a function of p. Equation (5.3) captures the collision probability that each packet transmission of the tagged node encounters if at least one of the other $N - 1$ nodes transmits in the same back-off time slot. In literature, there are mainly two ways to approximate τ: (1) $\tau = \frac{E[M_0]}{CW_2}$ [24], where $E[M_0] = \frac{1-p^{M_L+1}}{1-p}$ is the average number of transmission attempts each node made before the packet is successfully transmitted or discarded due to the retransmission limit M_L; and (2) $\tau = \frac{1}{\overline{CW_1}}$ [25], where $\overline{CW_1} = \frac{1-p-p(2p)^{M_b}}{1-2p} \frac{CW}{2}$ is the average back-off contention window size between two consecutive packet transmission attempts of the tagged node. Both approximations of τ can be substituted into (5.3) for solving p with certain N.

Since variables p and N are correlated in (5.3), a high-degree nonlinear equation whose computational complexity gets higher with an increase of degree N, the saturation throughput S_1, as a function of both variables, can only be evaluated by solving (5.3) numerically. Thus, the throughput model of (5.2) and (5.3) is a nonlinear system that does not provide a closed-form expression for S_1. Based on this numerical performance model, it is computational complex, by referring to numerical techniques, e.g., Newton's method [16] and fixed-point theorem [26], to conduct a performance comparison between IEEE 802.11 DCF and the other MAC candidate. Therefore, we aim to make some approximation on (5.3) to get an explicit closed-form relation between p and N, which can be directly substituted into (5.2) to simplify S_1 as a closed-form function of only N.

Some approximations are available in literature to simplify (5.3) (e.g., first-order approximation [25], asymptotic analysis [26]). However, as shown in Fig. 5.2, the accuracy of these approximations drops when N becomes larger. In [16], the exact relationship between p and N is depicted by solving (5.3) for p over a wide range of N using numerical techniques. It is stated that p increases both monotonically and logarithmically with N, provided that M_b, M_L, and CW are specified based on the IEEE 802.11b standard. Thus, to get a more accurate closed-form function between p and N, we use a *nonlinear least-squares curve-fitting method* to fit the relation between both variables:

$$\min_{\mathbf{a}=(a_1,a_2)} ||a_1 + a_2 \ln(\mathbf{N}) - \mathbf{P}||_2^2$$

$$subject\ to\ a_2 \geq 0 \tag{5.4}$$

where vectors $\mathbf{N} = \{n \mid n \in \mathbb{Z}^+\}$ and $\mathbf{P} = \{p_n \mid n \in \mathbb{Z}^+\}$ are data sets of N and p, respectively, satisfying (5.3), and $\mathbf{a} = (a_1, a_2)$ is the vector of the fitting coefficients. In (5.4), the bounded constraint makes the optimization problem converge fast to an optimal solution [27].

Proposition 6 *Global optimal fitting coefficients in (5.4) exist since the logarithmic nonlinear least-squares curve-fitting is a convex optimization problem.*

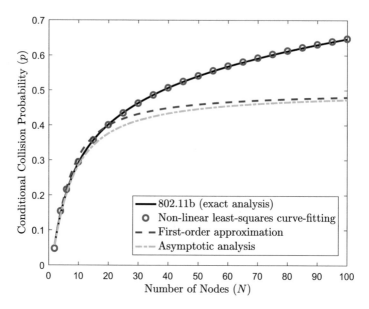

Fig. 5.2 Least-squares curve-fitting between p and N

The proof of Proposition 6 is given in Appendix H. The logarithmic curve-fitting relation obtained between p and N is

$$p \approx a_1 + a_2 \ln(N) = -0.0596 + 0.1534 \ln(N). \tag{5.5}$$

Figure 5.2 shows that the closed-form logarithmic fitting function is much more accurate to approximate p, over a wide range of N, than two existing approximations, i.e., first-order approximation in [25] and asymptotic analysis in [26]. Since the fitting function explicitly expresses p as a function of N, S_1 can be simplified as a closed-form function of only N, by substituting (5.5) into (5.2). However, since the average back-off contention window $\overline{CW_2}$ is a high-degree function of p, the approximate expression of S_1 is still complicated. The expression of $\overline{CW_2}$ can be approximated by an exponential function of p, because the Taylor expansion of an exponential function has a mathematical form similar to the expression of $\overline{CW_2}$. That is,

$$\overline{CW_2} = \frac{W_0}{2} + \frac{W_1}{2}p + \frac{W_2}{2}p^2 + \ldots + \frac{W_{M_b}}{2}p^{M_b} + \frac{W_{M_b}}{2}p^{M_b+1} + \ldots + \frac{W_{M_b}}{2}p^{M_L}$$

$$\approx b_1 + b_2 \exp(b_3 p)$$

$$= (b_1 + b_2) + b_2 b_3 p + b_2 \frac{b_3^2}{2!}p^2 + b_2 \frac{b_3^3}{3!}p^3 + \ldots$$

$$\tag{5.6}$$

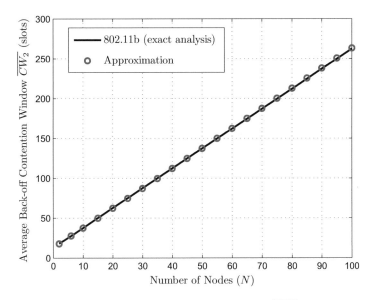

Fig. 5.3 An approximation of average back-off contention window $\overline{CW}_2$

where $(b_1, b_2, b_3) = (12.9590, 3.5405, 6.5834)$ are the coefficients of the exponential function obtained through the nonlinear least-squares curve-fitting method. Then, by substituting (5.5) into (5.6), $\overline{CW}_2$ can be further approximated by a closed-form function of N,

$$
\begin{aligned}
\overline{CW}_2 &\approx b_1 + b_2 \exp\left[b_3(a_1 + a_2 \ln(N))\right] \\
&= b_1 + b_2 \exp(b_3 a_1) \exp\left[b_3 a_2 \ln(N)\right].
\end{aligned}
\tag{5.7}
$$

Figure 5.3 shows that $\overline{CW}_2$ has a nearly linear relation with N since $b_3 a_2 \approx 1$ in (5.7).

By substituting (5.5) and (5.7) into (5.2), we obtain a simplified and closed-form expression of S_1, as a function of N, given by

$$
S_1(N) = \frac{N T_{pl}}{N T_s + \frac{N}{2} \frac{a_1 + a_2 \ln(N)}{1 - [a_1 + a_2 \ln(N)]} T_c + b_1 + b_2 \exp(b_3 a_1) \exp[b_3 a_2 \ln(N)]}.
\tag{5.8}
$$

where T_s, T_{pl}, and T_c are known parameters specified in the IEEE 802.11b standard. Figure 5.4 demonstrates that the simplified analytical function $S_1(N)$ is an accurate approximation of the numerical performance model [16, 23] represented by the nonlinear system of (5.2) and (5.3).

In the traffic saturation case, the average packet access delay (average packet service time), D_1, is defined as the duration from the instant that the packet becomes

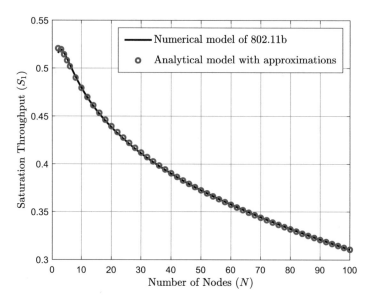

Fig. 5.4 Saturation throughput S_1 and its approximation

the head of the transmission queue to the instant that the packet is successfully transmitted, averaged over all transmitted packets of each node. Since D_1 is the denominator of S_1 [16], it can also be approximated by a closed-form analytical function of N, given by

$$D_1(N) = NT_s + \frac{N}{2} \frac{a_1 + a_2 \ln(N)}{1 - [a_1 + a_2 \ln(N)]} T_c + b_1 + b_2 \exp(b_3 a_1) \exp[b_3 a_2 \ln(N)].$$

$$(5.9)$$

Performance Analysis in a Traffic Non-saturation Case When the network is non-saturated, the average packet arrival rate λ of each node should not exceed its service capacity share μ_d. The packet queue at each node possibly becomes empty upon successful transmission of the previous packet. The derivation of the packet transmission probability should account for the fact that a node attempting a transmission only when it has packets to transmit. Thus, Eq. (5.3) should be revised to

$$p = 1 - (1 - \rho \cdot \tau)^{N-1} \qquad (5.10)$$

where $\rho = \frac{\lambda}{\mu_d}$ is the queue utilization ratio of an individual node, and $\rho\tau$ is the packet transmission probability of each node.

Due to its fairness property, the IEEE 802.11 system can be viewed as a server that schedules its resources to the contending nodes in a round robin manner [16]. In each scheduling cycle, every node (out of N nodes) can occupy an average fraction of $\frac{1}{N}$ system bandwidth to successfully transmit one packet. This service system is called processor sharing (PS) system. Thus, the IEEE 802.11 DCF can be modeled as an M/G/1/PS system with cumulative arrival rate $\lambda_s = N\lambda$ and service rate $\mu_s = N\mu_d$. According to [16], this M/G/1/PS system has the same access delay and queueing delay as the M/M/1 queueing system with equivalent average arrival rate λ_s and service rate μ_s. Thus, the average packet delay in the M/G/1/PS system is a summation of average packet access delay and average packet queueing delay (i.e., the duration from the instant that the packet arrives at the transmission queue to the instant that the packet becomes the queue head averaged over all transmitted packets of each node), given by

$$D_T = \frac{1}{\mu_s - \lambda_s} \tag{5.11}$$

where $\mu_s = \left[T_s + \frac{T_c}{2} + \frac{\overline{CW_2}}{N} \right]^{-1}$ [16].

Similar to the traffic saturation case, since p and N are correlated as in the high-degree nonlinear relation, (5.10) and (5.11) form a nonlinear system with two variables p and N that can be solved using numerical techniques [16, 24]. To get a simplified and closed-form performance expression as a function of N, one approach is to obtain an explicit relation between p and N. A first-order approximation of (5.10) and linearizing the transmission probability as $\tau \approx \frac{2CW}{(CW+1)^2}(1-p)$ [28] can be applied to simplify (5.10) to a quadratic equation of p, given by

$$p \approx (N-1)\lambda \left[N\left(T_s + \frac{T_c}{2} \right) + \overline{CW_2} \right] \tau$$

$$\approx (N-1)\lambda \left[\frac{2CWNT_s}{(CW+1)^2}(1-p) + \frac{NT_cCW}{(CW+1)^2}p + \frac{1}{1-p} \right], \quad (\tau, p \ll 1). \tag{5.12}$$

Then, with packet arrival rate λ, a closed-form relation between p and N can be established by solving the quadratic equation of p. However, this first-order approximation is accurate only when p and τ are much less than one for a small value of N, as shown in Fig. 5.5a, b. To have a more accurate approximation, we solve (5.10) for p over a wide range of N under the condition that all nodes are traffic non-saturated. It is found out that, with different values of λ, linearizing p as a function of N better characterizes the relation between p and N. Thus, a *linear least-squares curve-fitting method* is used to find a closed-form linear function between p and N, denoted by $p(N, \lambda)$ as an approximation of (5.10), shown in Fig. 5.5a, b. Substituting $p(N, \lambda)$ into $\overline{T_c}$ in (5.1) yields a closed-form function $\overline{T_c}(N, \lambda) = \frac{p(N,\lambda)}{1-p(N,\lambda)} T_c$.

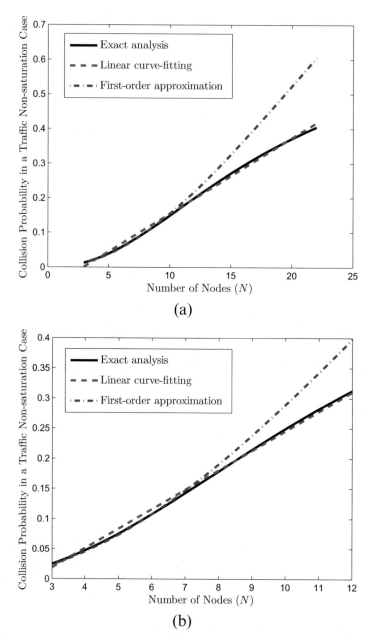

Fig. 5.5 Approximations for collision probability in a traffic non-saturation case. (**a**) $\lambda = 25$ packet/s. (**b**) $\lambda = 50$ packet/s

Since p shows a near linear relation with N for different values of λ and $\overline{CW_2}$ is approximately an exponential function of p in (5.6), $\overline{CW_2}$ can be approximately represented as an exponential function of N, denoted by $\overline{CW_2}(N, \lambda)$. Therefore, a closed-form expression for average packet delay D_T in terms of N is obtained as

$$D_2(N, \lambda) = \frac{1}{\mu_s(N, \lambda) - N\lambda} \tag{5.13}$$

where $\mu_s(N, \lambda) = \left[T_s + \frac{T_c(N,\lambda)}{2} + \frac{\overline{CW_2}(N,\lambda)}{N} \right]^{-1}$ is a closed-form expression for μ_s.

Similarly, the non-saturated network throughput, with the general form in (5.1), has the approximate and closed-form expression given as

$S_2(N, \lambda)$

$$= \frac{N T_{pl}}{N \left(T_s + \frac{T_c(N,\lambda)}{2} \right) + \overline{CW_2}(N, \lambda) + \left[1 - \frac{\lambda}{\mu_s(N,\lambda)} \right] \left[1 - (N - 1)\frac{\lambda}{\mu_s(N,\lambda)} \right] \left[\frac{1}{\lambda} - D_2(N, \lambda) \right]}. \tag{5.14}$$

Figures 5.6 and 5.7 show the exact values of average packet delay and non-saturation throughput as well as their accurate approximations over a wide range of N. It can be seen that, for λ equal to 25 and 50 packet/s, the traffic of each node

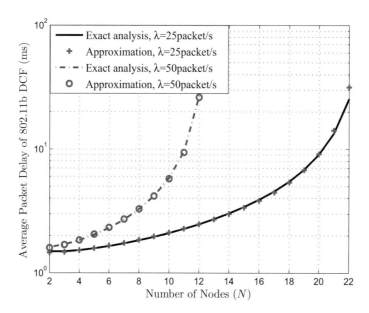

Fig. 5.6 Average packet delay of IEEE 802.11 DCF and its approximation for $\lambda = 25$ and 50 packet/s respectively

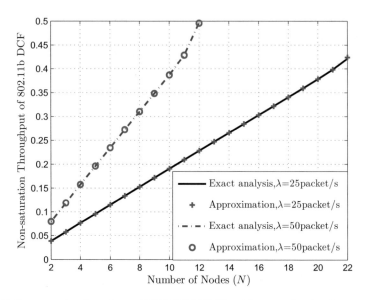

Fig. 5.7 Non-saturation throughput of IEEE 802.11 DCF and its approximation for $\lambda = 25$ and 50 packet/s respectively

enters the saturation state when N increases to the values greater than 22 and 12 respectively.

5.1.3.2   Closed-Form Performance Models for D-TDMA

Performance Analysis in A Traffic Saturation Case   When the network is saturated, we can obtain closed-form expressions of throughput and delay as a function of N. Since N in general varies slowly with respect to the frame duration, the network saturation throughput S_3 is approximately given by

$$S_3(N) = \frac{NT_{pl}}{NT_p + M_m T_m} \tag{5.15}$$

where T_{pl} is the duration of payload information of each data packet, T_p is the data packet duration including headers, M_m denotes the number of minislots in the control period of each D-TDMA frame, and T_m is the duration of each minislot.

Also, the average packet access delay of D-TDMA, denoted by D_3, can be expressed as

$$D_3(N) = NT_p + M_m T_m. \tag{5.16}$$

Performance Analysis in a Traffic Non-saturation Case In order to simplify the analysis of packet access delay and queueing delay, denoted by W_{st} and W_{qt} respectively, we assume that nodes release their data slots and randomly acquire new ones in the next frame, after transmitting a packet in the data transmission period of current frame [29]. This assumption guarantees that the service times of successive packets are i.i.d. random variables. Based on this assumption, the queue of each node in the traffic non-saturation case can be modeled as an M/G/1 queueing system [29], with an average service rate denoted by μ_t packet/s. We derive the distribution of packet service time W_{st} to calculate the average packet access delay, $E[W_{st}]$, in the M/G/1 system. Then, the P-K formula [30] can be used to calculate the average packet queueing delay, $E[W_{qt}]$, for each M/G/1 queue, based on the second moment of W_{st}, denoted by $E[W_{st}^2]$. As a result, the average packet delay D_4, which is the summation of $E[W_{st}]$ and $E[W_{qt}]$ (see Appendix I for the derivation of $E[W_{st}]$ and $E[W_{qt}]$), is given by

$$D_4 = E[W_{st}] + \frac{\lambda E[W_{st}^2]}{2[1 - \lambda E[W_{st}]]}. \tag{5.17}$$

Since $E[W_{st}]$ and $E[W_{st}^2]$ are both functions of N, D_4 is also a closed-form function of N, denoted by $D_4(N, \lambda)$.

As to the non-saturation throughput analysis, the probability that the queue of a tagged node is non-empty at the start of its designated time slot, denoted by P_{qn}, is given by

$$P_{qn} = \frac{\lambda}{\mu_t} \tag{5.18}$$

where $\mu_t = \frac{1}{E[W_{st}]} = \frac{2 - \lambda(M_c + N - 1)T_p}{(M_c + N + 1)T_p}$, M_c denotes the duration of each control period normalized to the unit of one D-TDMA data slot duration, according to the delay analysis in Appendix I.

Thus, we use random variable X to denote the number of nodes with non-empty queues at the start of their designated time slots during the time of one frame. The probability mass function (PMF) and the average of random variable X are given by [31]

$$P\{X = k\} = \binom{N}{k}\left(\frac{\lambda}{\mu_t}\right)^k \left(1 - \frac{\lambda}{\mu_t}\right)^{N-k}, \quad k = 0, 1, \ldots, N; \tag{5.19}$$

$$E[X] = N \cdot \frac{\lambda}{\mu_t}. \tag{5.20}$$

Hence, the network non-saturation throughput S_4 can be approximated as a function of N,

$$S_4(N, \lambda) = \frac{N\lambda T_{pl}}{\mu_t (NT_p + M_m T_m)}. \tag{5.21}$$

In summary, we derive simplified and closed-form throughput and delay expressions $S_1(N)$, $S_2(N, \lambda)$, $D_1(N)$, $D_2(N, \lambda)$ for the IEEE 802.11 DCF, and $S_3(N)$, $S_4(N, \lambda)$, $D_3(N)$, $D_4(N, \lambda)$ for D-TDMA, respectively, for both traffic saturation and non-saturation cases. The expressions can greatly simplify the MAC switching point calculation.

5.1.4   Adaptive MAC Solution

In this subsection, we present a MAC protocol which adapts to the changing traffic load in the MANET. The key element is to determine the MAC switching point, with which an appropriate candidate MAC protocol is selected to achieve better performance in terms of throughput and delay at each specific network traffic load condition. Based on the closed-form expressions derived in Sect. 5.1.3, we establish a unified performance analysis framework to evaluate the throughput and delay over a wide range of N for both non-saturated and saturated network traffic conditions. Taking throughput evaluation as an example, in this framework, when N is small, the network is non-saturated and the throughput is represented analytically by $S_2(N, \lambda)$ and $S_4(N, \lambda)$ for IEEE 802.11 DCF and D-TDMA, respectively. With an increase of N, packet service rates μ_d and μ_t of each node with both MAC protocols decrease consistently, making the queue utilization ratio of each node approach to one. After a specific network load saturation point in terms of N, say N_1 (N_2), where the arrival rate λ equals the service rate μ_d (μ_t), the network operating in IEEE 802.11 DCF (D-TDMA) enters the traffic saturation state. Thus, $S_1(N)$ and $S_3(N)$ are used to represent the network saturation throughput for each MAC candidate, respectively.

With this unified framework, performance comparison between the MAC candidates, with respect to N, can be conducted to calculate the MAC switching point. However, since IEEE 802.11 DCF and D-TDMA have different service capacity, the saturation points N_1 and N_2 are in general different, depending on λ. Therefore, the MAC switching point can be a specific network traffic load point, where the network with either IEEE 802.11 DCF or D-TDMA has four possible traffic state combinations: (1) the network is in the traffic saturation state with both MAC candidates; (2) the network is in the traffic non-saturation state with both MAC candidates; (3) the network is traffic saturated with IEEE 802.11 DCF and traffic non-saturated with D-TDMA; (4) the network is traffic non-saturated with IEEE 802.11 DCF and traffic saturated with D-TDMA. The established unified closed-form expressions facilitate performance comparison and the calculation of MAC switching point denoted by N_s (in terms of the number of nodes), for the four possible cases. The MAC switching point may vary, due to variations of λ at each node, in the homogeneous network traffic scenario. Algorithm 6 presents the detail steps of determining N_s. As an example, we illustrate step by step the switching point calculation for $\lambda = 25$ and 50 packet/s, respectively, based on network throughput comparison. Then, the complete MAC switching point calculation algorithm is provided considering all the possible cases.

Algorithm 6: MAC switching point calculation algorithm

Input   : The saturation points, N_1 and N_2, for IEEE 802.11 DCF and D-TDMA.
Output: The MAC switching point N_s.

1   **if** $N_1 < N_2$ **then**
2   **if** $S_1(N_1) > S_4(N_1, \lambda)$ **then**
3   **if** $S_1(N_2) < S_3(N_2)$ **then**
4               $N_s \leftarrow solving\ S_1(N) = S_4(N, \lambda)$;
5   **else**
6               $N_s \leftarrow solving\ S_1(N) = S_3(N)$;
7   **end**
8   **else if** $S_1(N_1) < S_4(N_1, \lambda)$ **then**
9           $N_s \leftarrow solving\ S_2(N, \lambda) = S_4(N, \lambda)$;
10   **else**
11          $N_s \leftarrow N_1$;
12   **end**
13 **else if** $N_1 > N_2$ **then**
14   **if** $S_3(N_2) > S_2(N_2, \lambda)$ **then**
15          $N_s \leftarrow solving\ S_2(N, \lambda) = S_4(N, \lambda)$;
16   **else if** $S_3(N_2) < S_2(N_2, \lambda)$ **then**
17   **if** $S_3(N_1) > S_1(N_1)$ **then**
18              $N_s \leftarrow solving\ S_2(N, \lambda) = S_3(N)$;
19   **else**
20              $N_s \leftarrow solving\ S_1(N) = S_3(N)$;
21   **end**
22   **else**
23          $N_s \leftarrow N_2$;
24   **end**
25 **else**
26   **if** $S_1(N_1) \geq S_3(N_1)$ **then**
27          $N_s \leftarrow solving\ S_1(N) = S_3(N)$;
28   **else**
29          $N_s \leftarrow solving\ S_2(N, \lambda) = S_4(N, \lambda)$;
30   **end**
31 **end**

---

1. $\lambda = 25$ packet/s:

    Step 1.   Compare the saturation points, N_1 and N_2, for both MAC candidates, $N_1 < N_2$;

    Step 2.   Compare the throughput of both MAC candidates at N_1, $S_1(N_1) > S_4(N_1, \lambda)$;

Step 3.   Compare the throughput of both MAC candidates at N_2, $S_3(N_2) > S_1(N_2)$;

Step 4.   The MAC switching point is calculated by solving equation $S_1(N) = S_4(N, \lambda)$, where the network has saturated traffic operating in IEEE 802.11 DCF and non-saturated traffic operating in D-TDMA.

2. $\lambda = 50$ packet/s:

Step 1.   Compare the saturation points N_1 and N_2 for both MAC candidates, $N_1 = N_2 = N^*$;

Step 2.   Compare the throughput of both MAC candidates at N^*, $S_1(N^*) = S_3(N^*)$;

Step 3.   The MAC switching point is obtained as $N_s = N^*$, where the network has saturated traffic operating in either candidate MAC protocol.

The MAC switching point can also be determined based on comparison of average packet delay between the MAC candidates, which is expected to generate similar results since a higher throughput corresponds to a lower packet delay. In theory, the average packet delay can be evaluated only when the packet arrival rate is less than the service rate, where the network traffic is in the non-saturation state. Otherwise, the packet delay will theoretically approach infinity. Therefore, when the MAC switching point locates at an N value where the network is in a traffic saturation state with either candidate MAC protocol, the average packet access delay, $D_1(N)$ and $D_3(N)$, can be used to calculate the switching point.

Due to node mobility, the number of nodes, N, may fluctuate around the switching point when nodes move relatively fast, resulting in undesired frequent MAC switching (taking account of switching cost). In order to benefit from the MAC switching, the performance gain should be higher than the switching cost. Therefore, the MAC switching point can be replaced by a switching interval. The MAC switching is triggered only if the number of nodes, N, varies beyond the switching interval. The length of the switching interval depends on the performance gain and switching cost. The switching cost can be calculated as the communication overhead consumed for periodic control information exchange among nodes to acquire the updated network traffic load information for distributed MAC switching. For a network with higher node mobility, nodes require more frequent control information exchange, resulting in an increased switching cost and a longer switching interval. Therefore, how to determine the optimal length of the switching interval to maximize the performance gain with the consideration of the switching cost and different node mobility patterns can be investigated in future research.

5.1.5 Numerical Results

In this subsection, we present analytical and simulation results for performance evaluation of both MAC candidates and the MAC switching point. The simulation results are used to demonstrate the accuracy of the MAC switching point calculation based on the closed-form expressions in Sect. 5.1.3. With an error-free wireless channel in the system, we use the network simulator, OMNeT++ [32], to simulate the IEEE 802.11b DCF and the D-TDMA. In the simulation, a fully-connected network over a $50\,m \times 50\,m$ square coverage area is deployed, where nodes are randomly scattered. Traffic arrivals for each node are realized as a Poisson process with λ being 25 and 50 packet/s, respectively, for the non-saturated traffic case, and with λ set as 500 packet/s for the saturated traffic case. The reason of using the same traffic model in computer simulations is to verify the accuracy of the analytical models proposed for the IEEE 802.11 DCF and the D-TDMA, since several assumptions and simplifications are made in the mathematical modeling and analysis, especially for traffic non-saturation conditions for both MAC schemes. For the IEEE 802.11 DCF, with homogeneous Poisson traffic arrivals, the 802.11 service system can be approximately modeled as an M/G/1/processor sharing (PS) system, for which the average packet delay analysis is greatly simplified; For the D-TDMA, to model the queue of each node in a traffic non-saturation condition as an M/G/1 queue, it is assumed that nodes release their data slots and randomly acquire new ones in the subsequent frame once the packet transmission is completed in current frame. Also, to simplify the derivation of the packet service time distribution, we normalize the control period of each D-TDMA frame to an integer multiple of one D-TDMA data slot duration and discretize the packet service time in the unit of one data slot, while neglecting the possibility of head-of-line (HOL) packets arriving within the duration of each data slot. These assumptions are necessary for a tractable analysis, and their effects on analysis accuracy should be evaluated by the simulations without the assumptions under the same traffic model. Each simulation point provides the average value of the corresponding performance metrics (i.e., throughput and delay). We also plot the 95% confidence intevals for each simulation result. Note that some confidence intervals are very small in the figures. Other main simulation parameters are summarized in Table 5.1.

5.1.5.1   Traffic Saturation Case

First, the saturation throughputs of both MAC candidates are plotted in Fig. 5.8a–c, for $M_m = 15, 25, 35$, respectively. It is observed that the analytical and simulation results closely agree with each other. As M_m increases, the saturation throughput of D-TDMA decreases since the length of control period in each D-TDMA frame increases, which reduces the channel utilization. The two MAC candidates have near opposite throughput variation trends as the network traffic load increases. For IEEE 802.11 DCF, the saturation throughput decreases with an increase of the traffic load. On the other hand, the saturation throughput of D-TDMA experiences a consistent rise when the number of nodes increases. Therefore, the two throughput curves intersect at a specific network traffic load value, for example $N = 12.5$ when $M_m = 35$. Before this value, IEEE 802.11 DCF outperforms D-TDMA and, after

Table 5.1 Simulation parameters used in IEEE 802.11b [33] and D-TDMA

MAC schemes		
Parameters	IEEE 802.11b	D-TDMA
Channel capacity	11 Mbps	11 Mbps
Basic rate	1 Mbps	1 Mbps
Back-off slot time	20 μs	–
Minimum contention window size (CW)	32	–
Maximum contention window size (W_m)	1024	–
Retransmission limit (M_L)	7	–
Guard time (GT) [19]	–	1 μs
PLCP & Preamble	192 μs	192 μs
MAC header duration	24.7 μs	24.7 μs
Packet payload duration (T_{pl})	$\frac{8184}{11}$ μs	$\frac{8184}{11}$ μs
Short interframe space (SIFS)	10 μs	–
ACK	10.2 μs	–
Distributed interframe space (DIFS)	50 μs	–
Minislot duration (T_m)	–	219.4 μs
Network size upper limit (M_m)	15/25/35 nodes	15/25/35 nodes
Queue length	10000 packets	10000 packets

this value, the D-TDMA performs better. Thus, the MAC switching point is the first integer number of nodes after the intersection, i.e., $N_s = 13$.

The average packet access delay of both MAC candidates in a traffic saturation case are plotted in Fig. 5.9a–c, for $M_m = 15, 25, 35$, respectively. It is observed that the MAC switching point is almost the same as that based on the saturation throughput.

5.1.5.2   Traffic Non-saturation Case

Figure 5.10a, b show how the network throughput changes with the number of nodes for both MAC schemes at $\lambda = 25$ and 50 packet/s, respectively. Again, the analytical results closely match the simulation results. In the simulation, we start from $N = 2$ where each node has a non-saturated traffic for both MAC candidates, and gradually increase the N value to $N = 35$. As N increases, the service rate for each node decreases, and the traffic at each node becomes saturated after N increases to a certain value. For IEEE 802.11 DCF, the saturation point locates at $N_1 = 23$ and 13 for λ equal to 25 packet/s and 50 packet/s, respectively. On the other hand, the corresponding saturation point of D-TDMA is $N_2 = 33$ and 13, respectively. When the traffic load is low, the non-saturation network throughput of IEEE 802.11 DCF is greater than that of D-TDMA and, therefore, nodes should choose IEEE 802.11 DCF as the initial MAC scheme. For $\lambda = 25$ packet/s, the MAC switching point is $N_s = 26$, where nodes with IEEE 802.11 DCF are traffic saturated. For $\lambda = 50$

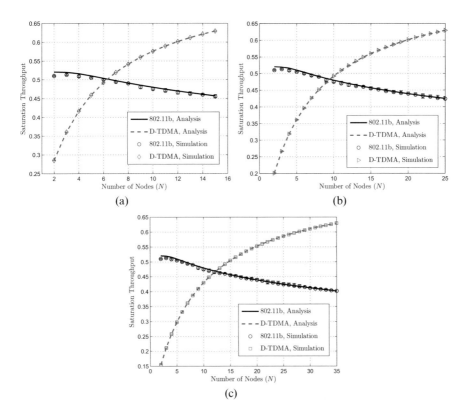

Fig. 5.8 Saturation throughput of both MAC schemes. (**a**) $M_m = 15$. (**b**) $M_m = 25$. (**c**) $M_m = 35$

packet/s, the switching point appears at $N_s = 13$, from which nodes with either MAC scheme have saturated traffic.

Figure 5.11a, b show the average packet delay for both MAC candidates in a traffic non-saturation case with $\lambda = 25$ and 50 packet/s respectively, for N varying from 2 to the largest integer within traffic non-saturation load region. We can see that the analytical results closely match the simulation results and the confidence intervals are very small. For $\lambda = 25$ packet/s, the two delay curves are expected to intersect at the network load point where IEEE 802.11 DCF becomes traffic saturated and D-TDMA is still traffic non-saturated. Thus, the MAC switching point exists as the saturation point of IEEE 802.11 DCF, denoted as $N_s = 23$. For $\lambda = 50$ packet/s, the two delay curves do not intersect in the traffic non-saturation state. Thus, the MAC switching point is expected to exist at a traffic saturated load point greater than $N = 12$, which can be obtained analytically as $N_s = 13$ based on comparison of average packet access delay for both MAC candidates, shown in Fig. 5.9c. The MAC switching point based on packet delay comparison is almost the same as that based on throughput comparison.

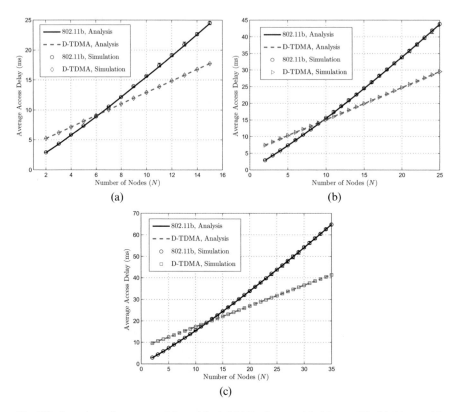

Fig. 5.9 Average packet access delay of both MAC schemes. (**a**) $M_m = 15$. (**b**) $M_m = 25$. (**c**) $M_m = 35$

5.2   Distributed and Service-Oriented MAC with Heterogeneous Traffic

In this section, we propose a distributed and adaptive hybrid MAC (DAH-MAC) scheme for a single-hop IoT-enabled MANET supporting voice and data services. A hybrid superframe structure is designed to accommodate packet transmissions from a varying number of mobile nodes generating either delay-sensitive voice traffic or best-effort data traffic. Within each superframe, voice nodes with packets to transmit access the channel in a contention-free period using distributed TDMA, while data nodes contend for channel access in a contention period using truncated CSMA/CA (T-CSMA/CA). In the contention-free period, by adaptively allocating time slots according to instantaneous voice traffic load, the MAC exploits voice traffic multiplexing to increase the voice capacity. In the contention period, a throughput optimization framework is proposed for the DAH-MAC, which maximizes the aggregate data throughput by adjusting the optimal contention window (CW) size according to voice and data traffic load variations. Numerical results show that the

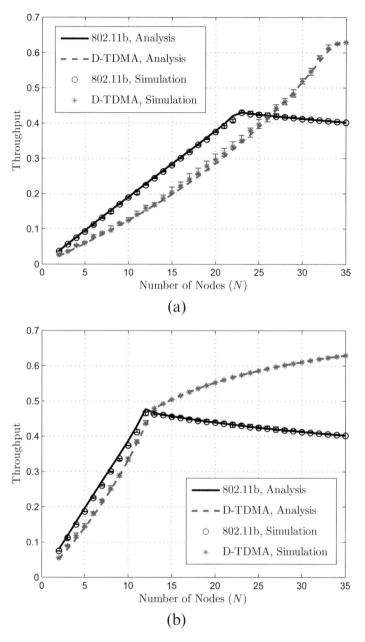

Fig. 5.10 Network throughput versus the number of nodes. (**a**) $\lambda = 25$ packet/s. (**b**) $\lambda = 50$ packet/s

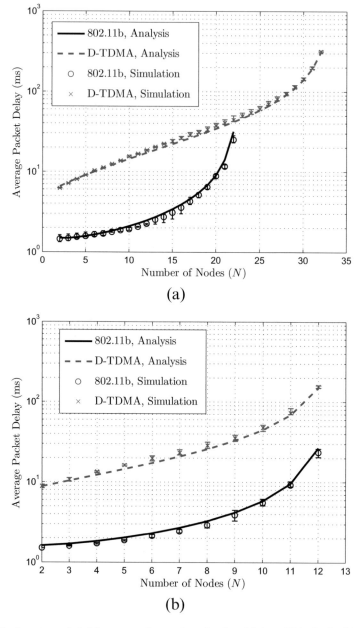

Fig. 5.11 Average packet delay versus the number of nodes. (a) $\lambda = 25$ packet/s. (b) $\lambda = 50$ packet/s

proposed MAC scheme outperforms existing QoS-aware MAC schemes for voice and data traffic in the presence of heterogeneous traffic load dynamics.

5.2.1 System Model

Consider a single-channel fully connected MANET [10, 13, 15], where each node can receive packet transmissions from any other node. The fully connected network scenario can be found in various MANET applications, including office networking in a building or in a university library where users are restricted to move in certain geographical areas [10], users within close proximity are networked with ad hoc mode in a conference site [8], M2M communications in a residential network for a typical IoT-based smart home application where home appliances are normally within the communication range of each other [2]. The channel is assumed error-free, and packet collisions occur when more than one node simultaneously initiate packet transmission attempts. Without any network infrastructure or centralized controller, nodes exchange local information with each other and make their transmission decisions in a distributed manner. The network has two types of nodes, voice nodes and data nodes, generating delay-sensitive voice traffic and best-effort data traffic, respectively. Each node is identified by its MAC address and a unique node identifier (ID) that can be randomly selected and included in each transmitted packet [34]. We use N_v and N_d to denote the total numbers of voice and data nodes in the network coverage area, respectively. Nodes are mobile with a low speed, making N_v and N_d change with time.

For delay-sensitive voice traffic, each packet should be successfully transmitted within a bounded delay to achieve an acceptable voice communications quality; otherwise, the packet will be dropped. Therefore, as a main QoS metric for voice traffic, packet loss rate should be carefully controlled under a given threshold, denoted by P_L (e.g., 10^{-2}). The generic *on/off* characteristic of voice traffic allows traffic multiplexing in transmission. Each voice source node is represented by an *on/off* model, which is a two-state Markov process with the *on* and *off* states being the talk spurt and silent periods, respectively. Both periods are independent and exponentially distributed with respective mean $\frac{1}{\alpha}$ and $\frac{1}{\beta}$. During a talk spurt, voice packets are generated at a constant rate, λ_v packet/s. As for best-effort data traffic, data nodes are expected to exploit limited wireless resources to achieve as high as possible aggregate throughput. It is assumed that each data node always has packets to transmit. Nodes in the network are synchronized in time, which can be achieved such as by using the 1PPS signal with a global positioning system (GPS) receiver [15, 34].

In the network, time is partitioned into superframes of constant duration, denoted by T_{SF}, which is set to have the same duration as the delay bound of voice traffic. Each superframe is further divided into three periods: control period (CTP), contention-free period (CFP) and contention period (CP), the durations of which are denoted by T_{ctrl}, T_{cfp} and T_{cp} respectively, as shown in Fig. 5.12. The control period

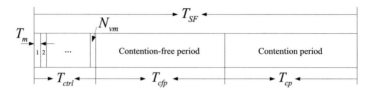

Fig. 5.12 Superframe structure

consists of N_{vm} fixed-duration (T_m) minislots, each with a unique minislot sequence number. It is to support a varying number of voice nodes in the network. Each voice node selects a unique minislot and broadcasts local information in its selected minislot, for distributed TDMA time slot allocation in the following contention-free period [19]. In the context of higher service priority to voice traffic, to avoid a complete deprivation of data service, there is a maximum fraction of time, φ (< 1), for voice traffic in each superframe. The value of φ is assumed known to all nodes when the network operation starts, and can be broadcast by the existing nodes in each control period. The voice capacity is the maximum number of voice nodes allowed in the network, denoted by N_{vm} (same as the number of minislots in each CTP), under the QoS constraint, which depends on φ. The period following the control period is the CFP, which is further divided into multiple equal-duration TDMA time slots, each slot having a unique sequence number. Each voice node with packets to transmit (referred to as active voice node) occupies one time slot to transmit a number of voice packets, called a *voice burst*.[3] Thus, the number of TDMA slots in the CFP is determined by the number of voice burst transmissions scheduled for the superframe, denoted by N_s ($N_s \leq N_v$).

The last period CP is dedicated to best-effort data nodes for transmission according to T-CSMA/CA. Data packet transmissions are based on CSMA/CA and are periodically interrupted by the presence of CTP and CFP.

5.2.2   The DAH-MAC Scheme

In the following, we illustrate how voice nodes access their TDMA slots in each CFP without a central controller. For data nodes accessing the channel using T-CSMA/CA, we highlight differences between the T-CSMA/CA within the proposed hybrid superframe structure and the traditional CSMA/CA.

---

[3] A voice burst is the packets generated by an active voice node within one superframe that can be transmitted over a time slot.

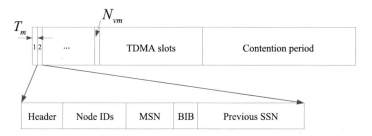

Fig. 5.13   Format of control packet broadcast in each minislot

5.2.2.1   Accessing Minislots

In the distributed MAC, each voice node needs to exchange information with neighboring voice nodes by broadcasting control packets in the minislots in the control period of each superframe. When the network operation starts, the number of minislots (voice capacity N_{vm}) in the control period should be determined in a distributed way under the constraint that voice packet loss rate is bounded by P_L and the summation of T_{ctrl} and T_{cfp} does not exceed $\varphi \cdot T_{SF}$ in each superframe. After N_{vm} is determined, each voice node randomly chooses one minislot in the CTP of a superframe, and broadcasts a control packet in its selected minislot [19]. Each node broadcasts its control packet in the same occupied minislot of each subsequent superframe,[4] until it is powered off or departs from the network. A control packet, shown in Fig. 5.13, includes five fields: a header, a set of IDs of the node's neighbors including the node itself, the node's occupied minislot sequence number (MSN, chosen from 1 to N_{vm}), buffer occupancy indication bit (BIB), the node's scheduled TDMA slot sequence number (SSN, a number within 0 to N_s) in the previous superframe.

Accessing a minislot from a tagged voice node is considered successful if the control packets received from other nodes in subsequent minislots contain the tagged node's ID [29]. Otherwise, an *access collision* happens due to simultaneous control packet transmissions in the same minislot by more than one node. All nodes involved in the collision wait until the next superframe to re-access one of the vacant minislots. The minislot accessing process is completed when all existing nodes successfully acquire their respective minislots. When a new node is powered on or entering the network coverage area, it first synchronizes in time with the start of a superframe, determines the number of minislots (based on φ), N_{vm}, and listens to all control packets in the CTP. Then, it randomly selects an unoccupied minislot and broadcasts a control packet in the minislot in the next superframe. If all N_{vm} minislots are occupied, which means the whole network reaches its voice capacity, the node defers its channel access and keeps sensing the CTPs

---

[4] To ensure fair minislot access, voice nodes re-select minislots after using the previous ones for a predefined number of successive superframes [35].

of subsequent superframes until some existing minislots are released due to node departures. After the minislot accessing is successful, the node keeps using the same minislot of subsequent superframes to broadcast its control packet.

5.2.2.2   Adaptive TDMA Slot Allocation

For efficient resource utilization, time slot allocation to voice nodes should adapt to traffic load variations. Taking account of the voice traffic *on/off* characteristic, only active nodes should be allocated one time slot each, in a superframe. We divide active voice nodes into two categories: Type I and Type II nodes. Type I nodes in the current superframe were not allocated a time slot in the previous superframe,[5] and are named "current-activated" nodes; Type II nodes remain active in both previous and current superframes, and are called "already-activated" nodes.

For each Type I node, voice traffic transits from the *off* state to the *on* state during previous superframe and generates voice packets before the node broadcasts a control packet in current superframe. Because of the randomness of state transition time from *off* to *on* in the previous superframe, packet transmissions from Type I nodes should have higher priority to be scheduled as early as possible according to their minislot accessing sequence, as long as Type II nodes can transmit within the delay bound, in order to minimize the possibility of Type I packet loss due to delay bound violation. Each Type II node has a time slot in the previous superframe and remains active in the current superframe. It should transmit packets no later than in the same time slot in the current superframe to meet the delay bound requirement. As an example, Fig. 5.14 illustrates how TDMA time slots are allocated in one superframe, with $N_v = 9$ and $N_{vm} = 10$, i.e., how to obtain current SSN based on the information in control packets. Each node has a unique minislot. Three types of important information in a broadcast control packet are shown: (1) MSN, the unique sequence number of a specific minislot; (2) BIB, which is 1 if the node has packets to transmit and 0 otherwise; (3) previous SSN, showing the sequence number of a TDMA time slot allocated to a voice node in the previous superframe, with SSN $= 0$ if the node was not allocated a TDMA time slot. Each entry in the left part of Fig. 5.14 discloses the information broadcast by the nodes in their minislots. The information broadcast by Type I and Type II active nodes is distinguished by the dashed-line and solid-line bounding rectangles, respectively. It can be seen that Nodes 8 and 5 are Type I with BIB $= 1$ and with previous SSN $= 0$, whereas Nodes 7, 1 and 4 are Type II with BIB $= 1$ and previous SSN $\neq 0$.

Based on the information provided by the control packets, the right part of Fig. 5.14 shows the time slot allocation in the current superframe. Node 8 accessing the first minislot among all the Type I nodes will transmit in the first time slot (with

---

[5] In most cases, the reason of not having a time slot is that the voice node has no packets to transmit. In some occasions when the instantaneous voice traffic load becomes heavy, an active node may not be able to get a time slot, resulting in packet dropping at the transmitter.

Node ID	MSN	BIB	Previous SSN	Current SSN
Node 7	1	1	3	3
Node 8	2	1	0	1
Node 5	3	1	0	4
Node 3	4	0	1	0
	5			
Node 1	6	1	2	2
Node 2	7	0	5	0
Node 6	8	0	4	0
Node 4	9	1	6	5
Node 9	10	0	0	0

Fig. 5.14 An example of TDMA time slot allocation

current SSN $= 1$) in the CFP. Node 5 is allocated a time slot after Nodes 1 and 7 because the latter two nodes are Type II nodes and should transmit packets no later than in their previously allocated time slots (with previous SSN $= 2$ and 3 respectively). Packet transmissions in current superframe from Node 4 (a Type II node with the largest previous SSN) are scheduled in a time slot with the index less than its previously allocated time slot.

5.2.2.3   T-CSMA/CA Based Contention Access

In DAH-MAC, best-effort data nodes access the channel within a CP of each superframe according to the T-CSMA/CA contention protocol, in which data nodes attempt packet transmissions according to CSMA/CA with exponential backoff [36] and the transmissions are periodically interrupted by the presence of a CTP and a CFP. Thus, the performance of T-CSMA/CA is different from the traditional CSMA/CA contention protocol without interruptions. First, the packet waiting time for transmission is increased by the interrupted periods; Second, before each source node initiates a packet transmission attempt at the end of its backoff counter decrementing process, it is required to check whether the remaining time in the CP is enough to support at least one packet transmission. To have an acceptable transmission attempt, the remaining time should be not less than the summation of a data packet duration (T_{pd}, including acknowledgment) and a gurad time (T_{gt}). This summation is called *conflict period*. If the remaining time in current CP is not long enough, a *virtual conflict* occurs with the imminent CTP of next superframe.

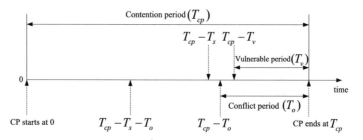

Fig. 5.15 An illustration of the CP

A *hold-on* strategy can be used to resolve the conflict [37, 38], in which the packet attempts are suspended until the start of next CP. Other nodes that are not involved in the conflict can still decrement their backoff counters within the conflict period until the end of current CP. When the next CP arrives, the transmission process resumes and the suspended packets are transmitted immediately after the channel is sensed idle for a distributed interframe space (DIFS).

By referring to some methods in [37], we give a detailed illustration inside the CP of each superframe, shown in Fig. 5.15, to highlight the differences between the T-CSMA/CA and the traditional CSMA/CA protocol. Suppose that the CP starts at time instant 0 and ends at T_{cp}. If a packet transmission attempt is initiated within the interval $\left[0, T_{cp} - T_s\right]$, the packet can be transmitted according to the CSMA/CA, either successfully or in collision, with a complete transmission duration T_s (T_{pd} plus a DIFS interval). The time instant $T_{cp} - T_o$ denotes the last time instant at which a packet transmission attempt can be initiated, and the conflict period is the following interval with duration T_o (T_{pd} plus T_{gt}), which is smaller than T_s. Thus, if a packet transmission starts in the interval $\left[T_{cp} - T_s, T_{cp} - T_o\right]$, the packet transmission time is on average $\frac{T_o + T_s}{2}$, assuming the transmission initiation instant is uniformly distributed within the interval. On the other hand, if the last packet transmission within the CP starts in the interval $\left[T_{cp} - T_s - T_o, T_{cp} - T_o\right]$, the transmission finishing point, denoted by $T_{cp} - T_v$, lies in the conflict period, where T_v is called *vulnerable period* [37] indicating the residual idle interval between the last transmission finishing point and the end of the CP; If no transmissions initiate during $\left[T_{cp} - T_s - T_o, T_{cp} - T_o\right]$, the starting point of the vulnerable period is $T_{cp} - T_o$, and the vulnerable period is the same as the conflict period. Thus, it can be seen that the vulnerable period is always not longer than the conflict period. The time interval $\left[0, T_{cp} - T_v\right]$ before the vulnerable period is called *non-vulnerable period*.

5.2.3 *Performance Analysis*

In this subsection, firstly, for a given maximum fraction of time (φ) for voice traffic in each superframe, the voice capacity, N_{vm}, under the packet loss rate bound is derived, which can facilitate voice session admission control. Secondly, for a specific N_v, the average number of voice burst transmissions scheduled in each superframe, $\overline{N_s}$, is derived, with which the average time duration of each CFP and CP, denoted by $\overline{T_{cfp}}$ and $\overline{T_{cp}}$, can be determined. Then, the aggregate throughput of the DAH-MAC for N_d data nodes is evaluated for each superframe, and maximized by adjusting the contention window size to the optimal value according to variations of N_v and N_d.

5.2.3.1 Voice Capacity

As mentioned in Sect. 5.2.2, when nodes come into the network coverage area, they distributedly calculate the number of minislots, N_{vm}, in each CTP, indicating the maximum number of voice nodes supported in the network. Thus, within the voice capacity region, the following inequality needs to be satisfied to guarantee that the time duration for voice traffic not exceed the maximum fraction (φ) of each superframe time,

$$T_{ctrl} + T_{cfpm} = N_{vm}T_m + N_{sm}\lceil B \rceil T_{pv} \leq \varphi T_{SF} \qquad (5.22)$$

where T_{cfpm} denotes the maximum value of T_{cfp}, T_m is the duration of each minislot, N_{sm} is the maximum value of N_s, indicating the maximum number of scheduled voice burst transmissions in a CFP to maintain the packet loss rate bound, B is the average size (number of voice packets) of a voice burst, $\lceil \rceil$ is the ceiling function, T_{pv} is the voice packet duration including header, and $\lceil B \rceil T_{pv}$ indicates the duration of each TDMA time slot (the duration of one time slot should allow an integer number of packet transmissions).

To determine N_{sm}, we first estimate that, with N_v voice nodes, how many generated voice packets in a superframe are required to be transmitted in the CFP to guarantee the packet loss rate bounded by P_L. Let X_i denote the number of packets generated by voice node i ($i = 1, 2, \ldots, N_v$) within a superframe, and y_m denote the maximum number of transmitted voice packets in the CFP to guarantee P_L. Since the length of a superframe is to be the same as the voice packet delay bound, lost packets are estimated as those generated but not transmitted within one superframe. Thus, y_m can be calculated by solving the following equation

$$\frac{E\left[X - y_m | X > y_m\right]}{E\left[X\right]} = P_L \qquad (5.23)$$

where $X = \sum_{i=1}^{N_v} X_i$.

Since $\{X_i, i = 1, 2, \ldots, N_v\}$ are independent and identically distributed (i.i.d) random variables, X can be approximated as a Gaussian random variable when N_v becomes relatively large (based on central limit theorem) [39], with mean $E[X]$ and variance $D[X]$ being $N_v E[X_i]$ and $N_v D[X_i]$ respectively. Thus, we estimate the distribution of X as a normal distribution, with which (5.23) is approximated as

$$\frac{\int_{y_m}^{N_v \cdot M_v} \frac{x - y_m}{\sqrt{2\pi N_v D[X_i]}} \cdot e^{-\frac{(x - N_v E[X_i])^2}{2N_v D[X_i]}} dx}{N_v E[X_i]} = P_L \tag{5.24}$$

where $M_v = \lambda_v \cdot T_{SF}$ denotes the maximum number of packets generated by a voice source node within one superframe.

In (5.24), to derive $E[X_i]$ and $D[X_i]$, we calculate the distribution of X_i, which is the probability of generating k packets by voice node i within a superframe ready for transmission in the CFP, denoted by $P(k)$.[6] According to the *on/off* source model, the probability of a voice node staying at *on* (*off*) state, denoted by P_{on} (P_{off}), at any time instant, is $\frac{\beta}{\alpha + \beta}$ ($\frac{\alpha}{\alpha + \beta}$). Let T_{on} (T_{off}) denote the time duration a voice node stays at *on* (*off*) state. We have

$$P(k) = P_{on} \cdot P\left\{\frac{k-1}{\lambda_v} < T_{on} \le \frac{k}{\lambda_v}\right\} + P_{off} \cdot P\left\{T_{SF} - \frac{k}{\lambda_v} < T_{off} \le T_{SF} - \frac{k-1}{\lambda_v}\right\}$$

$$= \frac{\beta}{\alpha + \beta}\left[e^{-\frac{\alpha(k-1)}{\lambda_v}} - e^{-\frac{\alpha k}{\lambda_v}}\right] + \frac{\alpha}{\alpha + \beta}\left[e^{-\beta\left(T_{SF} - \frac{k}{\lambda_v}\right)} - e^{-\beta\left(T_{SF} - \frac{k-1}{\lambda_v}\right)}\right],$$

$$(1 \le k \le M_v - 1) \tag{5.25}$$

$$P(M_v) = P_{on} \cdot P\left\{T_{on} > \frac{M_v - 1}{\lambda_v}\right\} + P_{off} \cdot P\left\{T_{off} \le \frac{1}{\lambda_v}\right\}$$

$$= \frac{\beta}{\alpha + \beta} e^{-\frac{\alpha(M_v - 1)}{\lambda_v}} + \frac{\alpha}{\alpha + \beta}\left(1 - e^{-\frac{\beta}{\lambda_v}}\right) \tag{5.26}$$

and

$$P(0) = 1 - \sum_{k=1}^{M} P(k). \tag{5.27}$$

With the probability distribution of X_i, the average voice burst size B can be obtained. Based on y_m and B, the maximum number of scheduled voice bursts, N_{sm}, in each CFP is derived. Then, with a specific φ, Algorithm 7 can be used to

---

[6] Since $\{X_i, i = 1, 2, \ldots, N_v\}$ have identical probability distribution, we simply drop the voice node index i.

determine the maximum number of voice nodes (voice capacity), N_{vm}, that can be supported in the network.

Algorithm 7: Voice capacity

Input : The maximum fraction of time, φ, for voice traffic in each superframe.
Output: Voice capacity N_{vm}, the CTP duration T_{ctrl}.

1 Initialization: $N_v \leftarrow 1$;
2 **do**
3      $y_m \leftarrow$ solving (5.24);
4      $N_{sm} \leftarrow \frac{y_m}{B}$;
5   **if** $N_v T_m + N_{sm} \lceil B \rceil T_{pv} \leq \varphi T_{SF}$ **then**
6 | $N_v \leftarrow N_v + 1$;
7   **else**
8         $N_{vm} \leftarrow N_v - 1$;
9         $T_{ctrl} \leftarrow N_{vm} T_m$;
10   **break**;
11   **end**
12 **while** $N_v > 0$;
13 **return** N_{vm} and T_{ctrl}.

5.2.3.2 Average Number of Scheduled Voice Bursts in a CFP

The actual number of generated voice bursts is likely less than N_{sm} and varies depending on the buffer occupancy states broadcast at the beginning of each superframe. In the following, for a specific N_v, we calculate the average number of scheduled voice bursts, $\overline{N_s}$, with which $\overline{T_{cfp}}$ and $\overline{T_{cp}}$ can be obtained.

We first determine the probability distribution of the number of active voice nodes, denoted by N_{av}, which broadcast control packets with BIB = 1 in their respective minislots of each superframe. Due to the voice source *on/off* characteristic, N_{av} is composed of two portions: (1) the number of nodes with a nonempty buffer staying at the *on* state, denoted by N_{av}^{on}; and (2) the number of nodes with a nonempty buffer staying at the *off* state, denoted by N_{av}^{off}.

Since packets periodically arrives at the transmission buffer every $\frac{1}{\lambda_v}$ second for each voice node at the *on* state, the probability of a voice node being active in its occupied minislot conditioned on that the node is at the *on* state can be derived by calculating a *posterior probability* as

$$P_{a|on} = P \left\{ \text{on at} \left(t_i - \frac{1}{\lambda_v} \right) \middle| \text{on at } t_i \right\} = e^{-\frac{\alpha}{\lambda_v}} \tag{5.28}$$

where t_i is the time instant that voice node i broadcasts in its selected minislot in the current superframe. Equation (5.28) indicates that the time duration of voice node i staying at the *on* state should last for at least the duration of $\frac{1}{\lambda_v}$ before it broadcasts at t_i to ensure a nonempty transmission buffer.

Similarly, the probability of voice node i being active at t_i, conditioned on that the node stays at the *off* state, is calculated as

$$P_{a|off} = P\left\{ \text{on at } (t_i - T)\,, \text{on at } \left(t_i - T + \frac{1}{\lambda_v}\right) \middle| \text{off at } t_i \right\}$$
$$= \frac{\beta}{\alpha}\left(e^{-\frac{\alpha}{\lambda_v}} - e^{-\alpha T}\right)$$

(5.29)

where $T = T_{SF} - T_{cfpm}$. Thus, $t_i - T$ is a calculation for the time instant of the end of the CFP in previous superframe.[7] Equation (5.29) indicates that voice node i should stay at the *on* state for at least the time interval of $\frac{1}{\lambda_v}$ after the end of previous CFP to ensure a nonempty transmission buffer before time instant t_i. Then, the probability distribution of N_{av}, denoted by $P_{N_{av}}(k)$, can be derived as

$$P_{N_{av}}(k)$$
$$= P\{N_{av}^{on} + N_{av}^{off} = k\}$$
$$= \sum_{i=0}^{N_v}\sum_{j\in\mathbf{A}} P\{N_{av}^{off} = k - N_{av}^{on} \,\middle|\, N_{av}^{on} = j, N^{on} = i\} P\{N_{av}^{on} = j \,\middle|\, N^{on} = i\} P\{N^{on} = i\}$$
$$= \sum_{i=0}^{N_v}\sum_{j\in\mathbf{A}} P(i, j, k) \qquad (0 \le k \le N_v)$$

(5.30)

where N^{on} denotes the number of nodes in the *on* state, set $\mathbf{A}$ denotes the value range of j depending on k and i, and

$$P(i, j, k) = \binom{N_v - i}{k - j} P_{a|off}^{k-j}\left(1 - P_{a|off}\right)^{N_v-i-k+j} \cdot \binom{i}{j} P_{a|on}^{j}\left(1 - P_{a|on}\right)^{i-j}$$
$$\cdot \binom{N_v}{i} P_{on}^{i} P_{off}^{N_v-i}.$$

(5.31)

---

[7] For calculation simplicity, we assume that the duration spent in current CTP before t_i is T_{ctrl}, and the duration of previous CFP is T_{cfpm}.

Then, the complete expression of $P_{N_{av}}(k)$ is obtained by delimiting j in (5.30) considering the following three cases:

1. $N_v - k > k$,

$$P_{N_{av}}(k) = \sum_{i=0}^{k}\sum_{j=0}^{i} P(i,j,k) + \sum_{i=k+1}^{N_v-k}\sum_{j=0}^{k} P(i,j,k) + \sum_{i=N_v-k+1}^{N_v}\sum_{j=i-N_v+k}^{k} P(i,j,k);$$

(5.32)

2. $N_v - k < k$,

$$P_{N_{av}}(k) = \sum_{i=0}^{N_v-k}\sum_{j=0}^{i} P(i,j,k) + \sum_{i=N_v-k+1}^{k}\sum_{j=i-N_v+k}^{i} P(i,j,k) + \sum_{i=k+1}^{N_v}\sum_{j=i-N_v+k}^{k} P(i,j,k);$$

(5.33)

3. $N_v - k = k$,

$$P_{N_{av}}(k) = \sum_{i=0}^{k}\sum_{j=0}^{i} P(i,j,k) + \sum_{i=k+1}^{N_v}\sum_{j=i-N_v+k}^{k} P(i,j,k).$$

(5.34)

Thus, with $P_{N_{av}}(k)$, the probability mass function (pmf) of the number of scheduled voice bursts, N_s, is given by

$$P_{N_s}(k) = \begin{cases} P_{N_{av}}(k) & (0 \le k \le N_{sm} - 1) \\ \displaystyle\sum_{r=N_{sm}}^{N_v} P_{N_{av}}(r) & (k = N_{sm}). \end{cases}$$

(5.35)

Finally, $\overline{N_s}$, $\overline{T_{cfp}}$, and $\overline{T_{cp}}$ can be obtained accordingly, based on $P_{N_s}(k)$.

5.2.3.3 A Data Throughput Optimization Framework for the DAH-MAC

In a CP, data nodes access the channel according to the T-CSMA/CA contention protocol. Since we evaluate the average aggregate throughput for data nodes in each superframe, $\overline{T_{cp}}$ is used to denote the average duration of a CP.[8] For the T-CSMA/CA, nodes are restricted to transmit packets within each CP. Before any transmission attempts, nodes are required to ensure that the remaining time in current CP is long enough for at least one packet transmission; otherwise, all transmission attempts are suspended until the next CP starts.

---

[8] All time intervals in a CP are normalized to the unit of an idle backoff time slot duration as commonly seen in CSMA/CA based systems.

According to the illustration inside the CP of each superframe in Fig. 5.15, the average value of T_v can be calculated by (5.36), assuming the transmission starting point is uniformly distributed inside $\left[\overline{T_{cp}} - T_s - T_o, \overline{T_{cp}} - T_o\right]$.

$$\overline{T_v} = (1 - \tau)^{N_d T_s} \cdot T_o + \left[1 - (1 - \tau)^{N_d T_s}\right] \cdot \frac{T_o}{2} = \frac{\left[1 + (1 - \tau)^{N_d T_s}\right] T_o}{2} \qquad (5.36)$$

where τ is the packet transmission probability of a data node with a nonempty transmission buffer at any backoff slot.

A generic time slot in the non-vulnerable period of the CP can be divided into three categories: (1) an idle backoff slot; (2) a complete transmission slot with the duration T_s (a successful transmission duration is assumed the same as a collision duration [40]); and (3) a restricted transmission slot with the duration $\frac{T_o + T_s}{2}$. Thus, the average duration of a generic time slot in the non-vulnerable period is calculated as

$$\sigma = (1 - \tau)^{N_d} + \left[1 - (1 - \tau)^{N_d}\right] T_a \qquad (5.37)$$

where $T_a = \left(\frac{T_s - T_o}{T_{cp} - T_o} \cdot \frac{T_s + T_o}{2} + \frac{\overline{T_{cp}} - T_s}{T_{cp} - T_o} \cdot T_s\right)$ denotes the average duration of a transmission slot in the non-vulnerable period.

Then, the probability that a generic slot is inside the vulnerable period is given by

$$p_v = \frac{\overline{T_v}}{\frac{\overline{T_{cp}} - \overline{T_v}}{\sigma} + \overline{T_v}}. \qquad (5.38)$$

Thus, the duration of a generic slot including the vulnerable period is derived as

$$\begin{aligned}
\sigma_d &= p_v + (1 - p_v)\sigma \\
&= p_v + (1 - p_v)\left[(1 - \tau)^{N_d} + \left[1 - (1 - \tau)^{N_d}\right] T_a\right].
\end{aligned} \qquad (5.39)$$

Since packet transmissions or collisions cannot happen in T_v, the packet collision probability for each node at any backoff slot in a traffic saturation case is expressed as

$$p = 1 - (1 - p_v)(1 - \tau)^{N_d - 1}. \qquad (5.40)$$

In (5.40), the transmission probability τ can be approximated, based on renewal reward theory, as a ratio of the average reward received during a renewal cycle over the average length of the renewal cycle [37, 40]. That is,

$$\tau = \frac{E[A]}{E[A] + E[W]} = \frac{\sum_{j=0}^{R_l-1} p^j}{\sum_{j=0}^{R_l-1} p^j + \sum_{j=0}^{R_l-1} \left(\frac{CW_j}{2} \cdot p^j\right)} \tag{5.41}$$

where $E[A]$ and $E[W]$ denote the average number of transmission attempts and backoff slots experienced, respectively, before a successful packet transmission; R_l is the retransmission limit; and $CW_j = 2^j CW (j = 0, 1, \ldots, M_b)$ is the contention window size in backoff stage j (CW is the minimum contention window size and M_b is the maximum backoff stage).

Therefore, the aggregate data saturation throughput[9] for the DAH-MAC is expressed as

$$\begin{aligned}
S_d &= \frac{N_d T_{pd} \tau (1-p)}{\sigma_d} \cdot \frac{\overline{T_{cp}}}{T_{SF}} \\
&= \frac{N_d T_{pd} \tau (1-p_v)(1-\tau)^{N_d-1}}{p_v + (1-p_v)[(1-\tau)^{N_d} + [1 - (1-\tau)^{N_d}] T_a]} \cdot \frac{\overline{T_{cp}}}{T_{SF}}
\end{aligned} \tag{5.42}$$

where T_{pd} is the data packet duration, and $\overline{T_{cp}}$ is a function of N_v.

From (5.42), we can see that when N_v and N_d are given and other system parameters are set, e.g., according to IEEE 802.11b standard [33], the saturation throughput S_d is a function of τ, and can be evaluated by solving (5.40) for τ numerically. We rewrite (5.42) as

$$S_d = \frac{T_{pd} \cdot \frac{\overline{T_{cp}}}{T_{SF}}}{\overline{T_{vd}}} \tag{5.43}$$

where $\overline{T_{vd}} = \frac{p_v + (1-p_v)[(1-\tau)^{N_d} + [1-(1-\tau)^{N_d}] T_a]}{N_d \tau (1-p_v)(1-\tau)^{N_d-1}}$. $\overline{T_{vd}}$ is called *average virtual transmission time*[41], which indicates the total average time experienced (including backoff waiting time, collision time and packet transmission time) in each CP to successfully transmit one packet.

Therefore, with different values of N_d, we evaluate the relationship between $\overline{T_{vd}}$ and τ, shown in Fig. 5.16. It can be seen that there exists an optimal transmission probability τ_{opt} that achieves a minimum of $\overline{T_{vd}}$. The reason of existing such an optimal transmission probability can be explained as follows: When $\tau < \tau_{opt}$, an

---

[9] The throughput in this section is normalized by the channel capacity.

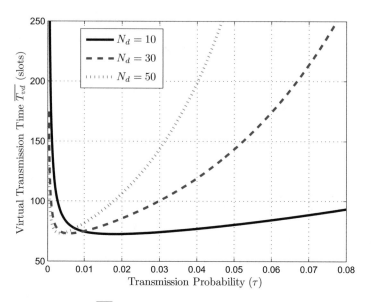

Fig. 5.16 The evaluation of $\overline{T_{vd}}$ in a function of τ ($\varphi = 0.5$, $N_v = 20$)

increasing amount of channel time remains idle before an transmission initiates, which consistently enlarges $\overline{T_{vd}}$ even if transmission collisions rarely happen in this scenario; However, when τ continues to increase beyond τ_{opt}, due to more frequent transmission attempts, the number of packet collisions rises, which consume an increasing fraction of channel time before packets are successfully transmitted. Therefore, the existence of τ_{opt} can be regarded as a compromise of the preceding two effects and achieves a minimum virtual transmission time and a maximum throughput. Since overheads consumed in one transmission collision are much greater than in an idle backoff slot, the optimal transmission probability is obtained as a relatively small value, as shown in Fig. 5.16, to lower the collision probability at the expense of consuming more idle slots. Therefore, our objective is to first derive τ_{opt} as a function of N_v and N_d. Then, by substituting τ_{opt} into (5.36)–(5.41), a closed-form mathematical relationship can be established between the optimal value of contention window size CW, denoted by CW_{opt}, and the heterogeneous network traffic load.

To do so, the expression of $\overline{T_{vd}}$ in (5.43) can be further derived as the summation of the following three terms:

$$\overline{T_{vd}} = \frac{p_v}{N_d(1-p_v)\tau(1-\tau)^{N_d-1}} + \frac{1-\tau}{N_d\tau} + \frac{\left[1-(1-\tau)^{N_d}\right]T_a}{N_d\tau(1-\tau)^{N_d-1}}. \qquad (5.44)$$

From (5.44), it is computational complex to obtain the first order derivative function of $\overline{T_{vd}}$ with respect to τ. The complexity mainly results from p_v which is a complex function of τ in (5.38). Thus, to make the derivation of $\overline{T_{vd}}$ tractable, an approximation of p_v can be obtained by simplifying $\overline{T_v}$, considering the following two cases:

1. For $\tau \geq 0.005$, since all time durations in a CP are normalized to the unit of an idle backoff slot duration, we have $T_s \gg 1$ (according to the IEEE 802.11b specification) and $N_d T_s \gg 1$. Thus,

$$\overline{T_v} = \frac{\left[1 + (1 - \tau)^{N_d T_s}\right] T_o}{2} \approx \frac{T_o}{2} \quad (\tau \geq 0.005); \tag{5.45}$$

2. For $\tau < 0.005$, the average duration of a generic time slot in the non-vulnerable period, σ, approaches 1. Moreover, since $\overline{T_v} \in \left[\frac{T_o}{2}, T_o\right]$, we have $\overline{T_v} \ll \overline{T_{cp}}$. Thus, Eq. (5.38) can be approximated as

$$p_v = \frac{\sigma \overline{T_v}}{\overline{T_{cp}} + (\sigma - 1)\overline{T_v}} \approx \frac{\sigma \overline{T_v}}{\overline{T_{cp}}} \approx \frac{\sigma T_o}{2\overline{T_{cp}}} \quad (\tau < 0.005). \tag{5.46}$$

As a result, by using $\frac{T_o}{2}$ (the lower bound of $\overline{T_v}$) to approximate $\overline{T_v}$, the approximation of p_v is

$$\tilde{p}_v = \frac{\frac{T_o}{2}}{\frac{\overline{T_{cp}} - \frac{T_o}{2}}{\sigma} + \frac{T_o}{2}}. \tag{5.47}$$

Therefore, by substituting (5.47) into (5.44) and after some algebraic manipulation, the approximation of $\overline{T_{vd}}$ is obtained as

$$\widetilde{T_{vd}} = \frac{\overline{T_{cp}}}{\overline{T_{cp}} - \frac{T_o}{2}} \cdot \frac{T_a - (T_a - 1)(1 - \tau)^{N_d}}{N_d \tau (1 - \tau)^{N_d - 1}}. \tag{5.48}$$

Then, by taking the first order derivative of $\widetilde{T_{vd}}$ with respect to τ and letting the derivative function equal to 0, we solve for an approximation of τ_{opt} (under the condition of $\tau \ll 1$) as a closed-form function of N_v and N_d, given by

$$\widetilde{\tau_{opt}} = \frac{\sqrt{1 + \frac{2(T_a - 1)(N_d - 1)}{N_d}} - 1}{(T_a - 1)(N_d - 1)}. \tag{5.49}$$

Figure 5.17 shows the accuracy of the approximation by plotting τ_{opt} and $\widetilde{\tau_{opt}}$ over a wide range of N_d. Note that although the optimal transmission probability, τ_{opt}, for each data node in a backoff slot decreases to a relatively small value with the increase of N_d, the probability of a successful packet transmission in a backoff slot is much higher than τ_{opt} when N_d becomes large, to achieve the maximized throughput.

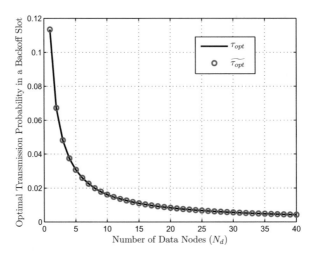

Fig. 5.17 Optimal transmission probability in each backoff slot for data nodes ($N_v = 20$, $\varphi = 0.5$)

Then, by substituting $\widetilde{\tau_{opt}}$ and $\widetilde{p}_v$ into (5.37)–(5.41), we derive an approximate expression for the optimal contention window CW_{opt}, as a closed-form function of N_v and N_d, given by

$$\widetilde{CW_{opt}} = \frac{\left(1 - \widetilde{\tau_{opt}}\right)\left(1 - \widetilde{p}^{R_l}\right)}{\widetilde{\tau_{opt}}\,(1 - \widetilde{p})\left(\sum_{j=0}^{M_b} 2^{j-1}\widetilde{p}^j + \sum_{j=M_b+1}^{R_l-1} 2^{M_b-1}\widetilde{p}^j\right)} \tag{5.50}$$

where $\widetilde{p} = 1 - (1 - \widetilde{p}_v)(1 - \widetilde{\tau_{opt}})^{N_d-1}$.

The proposed analytical framework not only provides an effective way to evaluate the performance of the DAH-MAC in supporting both voice and data traffic, but also provides some insights in MAC design in practical engineering for performance improvement: First, the voice capacity, N_{vm}, and the maximum number of voice bursts, N_{sm}, scheduled for each superframe are derived based on the analytical model and used as a reference for engineers in the protocol design to guarantee the voice delay bound in presence of the voice traffic load dynamics; Second, with the closed-form mathematical relationship provided in (5.50), data nodes operating T-CSMA/CA in each CP can adaptively adjust the minimum contention window size CW to the optimal value $\widetilde{CW_{opt}}$, based on the updated heterogeneous network traffic load information acquired in each superframe, to achieve consistently maximum aggregate data throughput.

5.2.4 Numerical Results

In this subsection, simulation results are provided to validate the accuracy of the analytical results. All simulations are carried out using OMNeT++ [32, 42, 43]. Nodes are interconnected and each source node randomly selects one of the rest nodes as its destination node. We run each simulation for 10000 superframe intervals to generate one simulation point. The main simulation settings for different MAC schemes with voice and data traffic are listed in Table 5.2. For a voice source, the GSM 6.10 codec is chosen for encoding the voice stream, with which voice packet payload size is 33 bytes and packets interarrival interval is 20 ms when the voice source node is at the *on* state [39]. Packet arrivals for each best-effort data node follow a Poisson process with the average arrival rate λ_d. We set λ_d to 500 packet/s to ensure each data transmission buffer is always saturated.

We first study the voice capacity, determined by Algorithm 7 in Sect. 5.2.3, with a variation of the maximum fraction of time (φ) for voice traffic in each superframe. Then, with a specific φ, the maximum number of voice burst transmissions supported in a CFP to guarantee the packet loss rate bound and the average number of scheduled voice bursts are both evaluated. For performance metrics, the voice packet loss rate and aggregate data throughput are considered. We also compare the performance of the proposed DAH-MAC scheme with two well-known MAC protocols.

5.2.4.1 Voice Capacity

The voice capacity in the network, with a variation of φ, for the DAH-MAC is plotted in Fig. 5.18. The analytical results are obtained according to Algorithm 7 in Sect. 5.2.3.1. It can be seen that the analytical results closely match the simulation results especially when the voice capacity region is relatively large, since using the central limit theorem to approximate the distribution of X in (5.23) becomes more accurate when N_v gets larger.

5.2.4.2 Number of Scheduled Voice Bursts (Time Slots) in a CFP

We also evaluate the number of voice burst transmissions (time slots) in a CFP with different N_v in the voice capacity region. Figure 5.19 shows the average number of scheduled voice bursts ($\overline{N_s}$) in each superframe. It can be seen that the analytical and simulation results closely match, which verifies the accuracy of our analysis. In Fig. 5.19, we also plot the maximum supported voice bursts (N_{sm}) in each CFP. Due to the randomness of voice packet arrivals, the instantaneous voice traffic load fluctuates on a per-superframe basis. The gap between N_{sm} and $\overline{N_s}$ indicates the number of time slots allocated to voice bursts in each CFP is commonly below the maximum allowable value. Therefore, by adapting to the realtime voice traffic load

Table 5.2 Simulation parameter settings [33, 39]

MAC schemes			
Parameters	DAH-MAC	Busy-tone contention protocol [39]	D-PRMA [15]
Channel capacity	11Mbps	11Mbps	11Mbps
Backoff slot time	20 μs	20 μs	n.a.
Minimum CW size (voice/data)	n.a.	8/32	n.a.
Maximum CW size (voice/data)	n.a.	16/1024	n.a.
Backoff stage limit (voice/data)	n.a./5	1/5	n.a.
Retransmission limit (voice/data)	n.a./7	2/7	n.a.
PLCP and preamble	192 μs	192 μs	192 μs
MAC header	24.7 μs	24.7 μs	24.7 μs
RTP/UDP/IP headers (voice)	$\frac{4\cdot8}{11}$ μs	$\frac{4\cdot8}{11}$ μs	$\frac{4\cdot8}{11}$ μs
Payload length (voice/data)	$\frac{33\cdot8}{11}$ / $\frac{8184}{11}$ μs	$\frac{33\cdot8}{11}$ / $\frac{8184}{11}$ μs	$\frac{33\cdot8}{11}$ / $\frac{8184}{11}$ μs
AIFS/DIFS (voice/data)	n.a./50 μs	30/50 μs	n.a.
Minislot duration	0.25 ms	n.a.	0.41 ms
Time slot duration	1.22 ms	n.a.	1.64 ms
Transmission time (voice/data)	0.244/1.18 ms	0.244/1.18 ms	0.244/1.18 ms
Gurad time (T_{gt})	20 μs	n.a.	n.a.
Average *on/off* time $\left(\frac{1}{\alpha}/\frac{1}{\beta}\right)$	352/650 ms	352/650 ms	352/650 ms
Minislot contention probability (voice/data)	n.a.	n.a.	0.6/0.2
Transmission queue length	10000 packets	10000 packets	10000 packets
Superframe time (delay bound)	100 ms	100 ms	100 ms

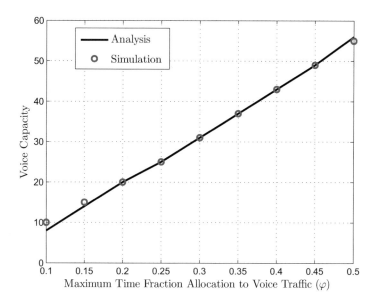

Fig. 5.18 Voice capacity region with different φ

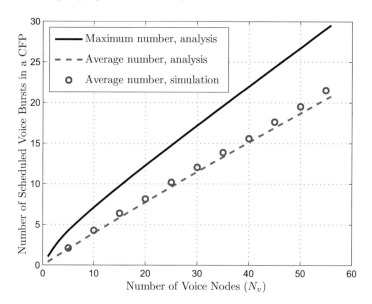

Fig. 5.19 Number of scheduled time slots for voice traffic ($\varphi = 0.5$)

in each superframe, the proposed distributed TDMA time slot allocation achieves a high resource utilization.

Figure 5.20 shows the average time allocated to voice and data traffic in each superframe with a specific φ. In Fig. 5.20a, we can see that the average time of each

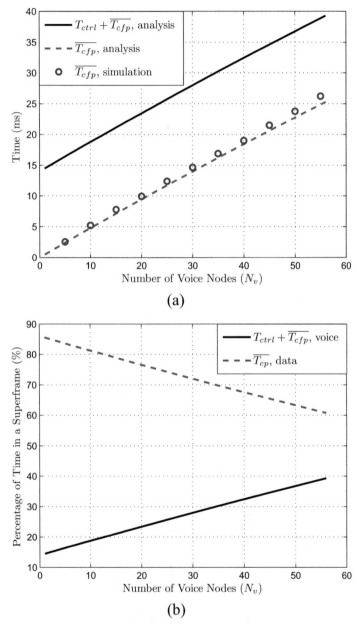

Fig. 5.20 Average time allocation in a superframe ($\varphi = 0.5$). (**a**) Durations of CTP and CFP for voice traffic. (**b**) Percentage of time for voice and data traffic

CFP ($\overline{T_{cfp}}$) for voice burst transmissions increases with N_v, with a fixed duration of each CTP (T_{ctrl}) for accommodating a certain number of voice nodes, which is determined according to the voice capacity with $\varphi = 0.5$; Fig. 5.20b shows the average percentage of time allocated to voice and data traffic in each superframe. It can be seen that within the capacity region, the average time allocated to voice traffic is always bounded by φT_{SF} under the packet loss rate constraint, and the residual average superframe time are occupied by data traffic.

5.2.4.3 Voice Packet Loss Rate

Packet loss rate for voice traffic in a CFP is evaluated with different φ in Fig. 5.21. It is observed that the simulation results are close to the analytical results. Although some performance fluctuations appear when N_v is relatively small due to the central limit theorem approximation and the rounding-off effect in deriving N_{sm} (set as a simulation parameter), the packet loss rate is always below the performance bound within the voice capacity region, which verifies the effectiveness of our proposed MAC in supporting voice service. If the number of minislots in the control period of each superframe is set beyond the voice capacity N_{vm}, the packet loss rate rises dramatically as shown in Fig. 5.21. Therefore, Algorithm 7 in Sect. 5.2.3.1 is employed in the DAH-MAC to calculate N_{vm} with different requirement of φ, which controls the number of voice nodes N_v within the capacity region to guarantee a bounded packet loss rate.

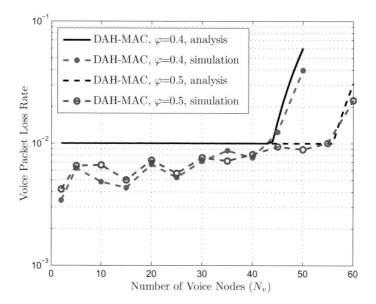

Fig. 5.21 Voice packet loss rate in a CFP with different φ

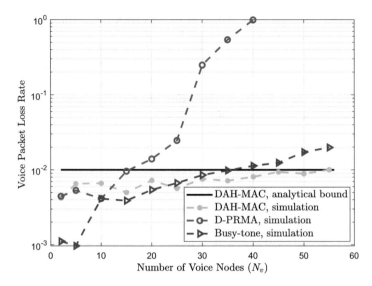

Fig. 5.22 A comparison of voice packet loss rates ($N_d = 10$, $\varphi = 0.5$)

Figure 5.22 displays a comparison of voice packet loss rates between the proposed DAH-MAC and two well-known MAC protocols: D-PRMA protocol [15] and busy-tone contention protocol [39], with a variation of N_v. The latter two MAC protocols are both effective in supporting voice packet transmissions. We can see that the D-PRMA can guarantee a bounded packet loss rate when N_v is relatively small. However, the packet loss rate increases dramatically since contention collisions rise when an increasing number of voice nodes start to contend for the transmission opportunity in each available time slot. Thus, the voice capacity region for D-PRMA is limited. Different from the D-PRMA, the busy-tone contention protocol grants a deterministic channel access priority for voice traffic. Thus, it can be seen that the voice packet loss rate is guaranteed over a wide range of N_v. Nevertheless, due to the contention nature, a consistent increase of voice packet collisions lead to accumulated channel access time, and the packet loss rate eventually exceeds the bound after around 35 voice nodes are admitted. In the DAH-MAC, the proposed distributed TDMA can admit more voice sessions by setting a higher value of φ, at the expense of local information exchanges in each enlarged control period. As can be seen in Fig. 5.22, the voice capacity region of the proposed MAC can be larger than the other two MAC protocols with a bounded packet loss rate.

5.2.4.4   Aggregate Best-Effort Data Throughput

We first evaluate the average channel utilization of T-CSMA/CA in each CP, defined as the ratio of average time used for successful data packet transmissions in a CP

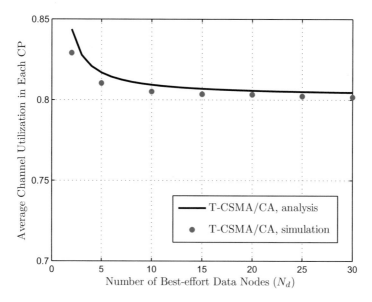

Fig. 5.23 Channel utilization for data traffic in each CP ($N_v = 20$, $\varphi = 0.5$)

to average duration of the CP. It can be seen in Fig. 5.23 that the T-CSMA/CA achieves consistently high channel utilization with variations of data traffic load, since the proposed throughput analytical framework maximizes the T-CSMA/CA channel utilization within the DAH-MAC superframe structure.

Then, we make a comparison of aggregate data throughput between the proposed DAH-MAC and the busy-tone contention protocol with a variation of N_d and under different voice traffic load conditions. To ensure a fair comparison, we set φ as 0.33 for DAH-MAC to achieve the same voice capacity ($N_{vm} = 35$) with the busy-tone contention protocol. First, when $N_v = 35$ representing a high voice traffic load condition, it can seen from Fig. 5.24a that the DAH-MAC can achieve a consistently higher data throughput than the busy-tone contention protocol. This is because the distributed TDMA can achieve better resource utilization in high voice traffic load conditions, and the aggregate data throughput is maximized over a wide range of N_d; for the busy-tone contention protocol, with a high voice traffic load, an increasing fraction of channel time is consumed for voice collisions resolution, and a large number of voice nodes also limit the channel access opportunity for data traffic, since voice traffic has absolute priority of accessing the channel. The data throughput comparison is also conducted when the voice traffic load is moderate. It can be seen in Fig. 5.24b that throughputs of the DAH-MAC and the busy-tone contention protocol experience a similar trend with the increase of N_d as in Fig. 5.24a, except for the cases where the busy-tone contention protocol achieves a higher throughput when N_d becomes relatively small. The advantage of busy-tone contention becomes more notable when the voice traffic load gets lower, as can be seen in Fig. 5.24c where $N_v = 5$. Therefore, the results from Fig. 5.24b,

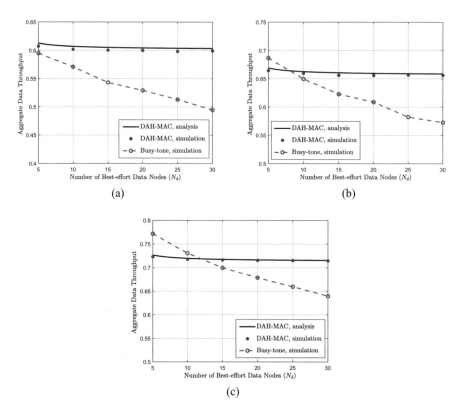

Fig. 5.24 A comparison of the DAH-MAC maximum data throughput ($\varphi = 0.33$) with the busy-tone contention protocol. (**a**) $N_v = 35$. (**b**) $N_v = 20$. (**c**) $N_v = 5$

c demonstrate some effectiveness of busy-tone based contention when N_v and N_d are relatively small, since in a low heterogeneous network traffic load condition, contention collisions among voice or data nodes are largely reduced, thus improving the channel utilization of the contention-based MAC protocol. Overall, over a wide range of N_d, our proposed MAC can achieve a consistently higher throughput especially in a high voice traffic load condition.

We further conduct the throughput comparison between the DAH-MAC and the D-PRMA. It is shown in Fig. 5.25 that the DAH-MAC can achieve both a larger voice capacity region ($N_{vm} = 35$ with $\varphi = 0.33$) and a higher data throughput than the D-PRMA. Since the D-PRMA is designed to support the QoS of voice traffic, the throughput for best-effort data traffic is suppressed because data nodes contend the channel with a lower probability and they can only transmit once in a slot upon the successful contention. Also, the D-PRMA uses slotted-Aloha based mechanism to contend for the transmission opportunity, which is inferior to the CSMA/CA based mechanism in terms of collisions resolution. We can see that the data throughput starts to decrease when N_d becomes large due to accumulated contention collisions.

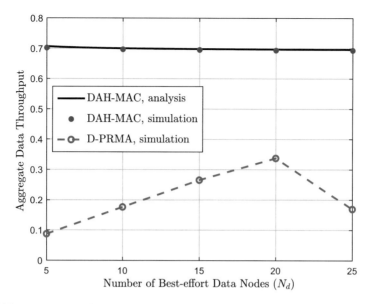

Fig. 5.25 A comparison of the DAH-MAC maximum data throughput ($\varphi = 0.33$, $N_v = 10$) with the D-PRMA protocol

5.3 Summary

In this chapter, we have presented adaptive MAC solutions for supporting both homogeneous and heterogeneous wireless traffic in IoT-enabled mobile networks. Considering a fully-connected MANET with homogeneous best-effort data traffic, we develop adaptive MAC between IEEE 802.11 and D-TDMA to maximize network performance over traffic load variations. An approximate and closed-form analytical model is established to calculate the optimal MAC switching point in terms of total number of nodes in the network, upon which nodes make switching decisions between the MAC candidates in a distributed manner when the network traffic load varies; For a MANET supporting heterogeneous services, we have proposed a distributed and traffic-adaptive hybrid MAC scheme, in which voice nodes are allocated time slots in a distributed way by adapting to their instantaneous transmission buffer states and data nodes contend to access the channel in a contention period of each superframe according to the T-CSMA/CA. The proposed hybrid MAC exploits the voice traffic multiplexing while guaranteeing a voice packet loss rate bound, and reduces the congestion level for data nodes. A data throughput analytical and optimization framework is developed, in which a closed-form mathematical relation is established between the MAC layer parameter (i.e., the optimal contention window size) and the number of voice and data nodes in the network. With this framework, the maximum aggregate data throughput can be achieved and be adaptive to variations of the heterogeneous network traffic load.

Extensive simulation results verify the effectiveness of the proposed adaptive MAC solutions in providing heterogeneous QoS guarantee and achieving adaptiveness to traffic load dynamics.

References

1. J. Gubbi, R. Buyya, S. Marusic, M. Palaniswami, Internet of Things (IoT): a vision, architectural elements, and future directions. Future Gener. Comput. Syst. **29**(7), 1645–1660 (2013)
2. A. Al-Fuqaha, M. Guizani, M. Mohammadi, M. Aledhari, M. Ayyash, Internet of Things: a survey on enabling technologies, protocols, and applications. IEEE Commun. Surv. Tutorials **17**(4), 2347–2376 (2015)
3. H. Nishiyama, T. Ngo, S. Oiyama, N. Kato, Relay by smart device: innovative communications for efficient information sharing among vehicles and pedestrians. IEEE Veh. Technol. Mag. **10**(4), 54–62 (2015)
4. J. Liu, N. Kato, A Markovian analysis for explicit probabilistic stopping-based information propagation in postdisaster ad hoc mobile networks. IEEE Trans. Wirel. Commun. **15**(1), 81–90 (2016)
5. J.-R. Cha, K.-C. Go, J.-H. Kim, W.-C. Park, TDMA-based multi-hop resource reservation protocol for real-time applications in tactical mobile adhoc network, in *Proc. IEEE MILCOM'10*, 2010, pp. 1936–1941
6. Q. Ye, W. Zhuang, Token-based adaptive MAC for a two-hop Internet-of-Things enabled MANET. IEEE Internet Things J. **4**(5), 1739–1753 (2017)
7. S. Tozlu, M. Senel, W. Mao, A. Keshavarzian, Wi-Fi enabled sensors for Internet of Things: a practical approach. IEEE Commun. Mag. **50**(6), 134–143 (2012)
8. M. Natkaniec, K. Kosek-Szott, S. Szott, G. Bianchi, A survey of medium access mechanisms for providing QoS in ad-hoc networks. IEEE Commun. Surv. Tutor. **15**(2), 592–620 (2013)
9. Z. Haas, J. Deng, Dual busy tone multiple access (DBTMA)-a multiple access control scheme for ad hoc networks. IEEE Trans. Commun. **50**(6), 975–985 (2002)
10. L. Lei, S. Cai, C. Luo, W. Cai, J. Zhou, A dynamic TDMA-based MAC protocol with QoS guarantees for fully connected ad hoc networks. Kluwer J. Telecommun. Syst. **60**, 43–53 (2014)
11. I. Chlamtac, M. Conti, J.J.-N. Liu, Mobile ad hoc networking: imperatives and challenges. Ad Hoc Netw. **1**(1), 13–64 (2003)
12. A. Abdrabou, W. Zhuang, Stochastic delay guarantees and statistical call admission control for IEEE 802.11 single-hop ad hoc networks. IEEE Trans. Wirel. Commun. **7**(10), 3972–3981 (2008)
13. H. Jiang, P. Wang, W. Zhuang, A distributed channel access scheme with guaranteed priority and enhanced fairness. IEEE Trans. Wirel. Commun. **6**(6), 2114–2125 (2007)
14. W. Hu, H. Yousefi'zadeh, X. Li, Load adaptive MAC: a hybrid MAC protocol for MIMO SDR MANETs. IEEE Trans. Wirel. Commun. **10**(11), 3924–3933 (2011)
15. S. Jiang, J. Rao, D. He, X. Ling, C.C. Ko, A simple distributed PRMA for MANETs. IEEE Trans. Veh. Technol. **51**(2), 293–305 (2002)
16. K. Medepalli, F. Tobagi, System centric and user centric queueing models for IEEE 802.11 based wireless LANs, in *Proc. IEEE BroadNets'05*, 2005, pp. 612–621
17. C. Doerr, M. Neufeld, J. Fifield, T. Weingart, D.C. Sicker, D. Grunwald, MultiMAC-an adaptive MAC framework for dynamic radio networking, in *Proc. IEEE DySPAN'05*, 2005, pp. 548–555
18. L. Kleinrock, F. Tobagi, Packet switching in radio channels: part I–carrier sense multiple-access modes and their throughput-delay characteristics. IEEE Trans. Commun. **23**(12), 1400–1416 (1975)

19. N. Wilson, R. Ganesh, K. Joseph, D. Raychaudhuri, Packet CDMA versus dynamic TDMA for multiple access in an integrated voice/data PCN. IEEE J. Sel. Areas Commun. **11**(6), 870–884 (1993)
20. A. Abdrabou, W. Zhuang, Service time approximation in IEEE 802.11 single-hop ad hoc networks. IEEE Trans. Wirel. Commun. **7**(1), 305–313 (2008)
21. G. Berger-Sabbatel, A. Duda, M. Heusse, F. Rousseau, Short-term fairness of 802.11 networks with several hosts, in *Mobile and Wireless Commun. Netw.*, 2005, pp. 263–274
22. G. Berger-Sabbatel, A. Duda, O. Gaudoin, M. Heusse, F. Rousseau, Fairness and its impact on delay in 802.11 networks, in *Proc. IEEE GLOBECOM'04*, vol. 5, 2004, pp. 2967–2973
23. K. Medepalli, F. Tobagi, Throughput analysis of IEEE 802.11 wireless LANs using an average cycle time approach, in *Proc. IEEE GLOBECOM'05*, vol. 5, 2005, pp. 3007–3011
24. L. Cai, X. Shen, J.W. Mark, L. Cai, Y. Xiao, Voice capacity analysis of WLAN with unbalanced traffic. IEEE Trans. Veh. Technol. **55**(3), 752–761 (2006)
25. Y. Tay, K.C. Chua, A capacity analysis for the IEEE 802.11 MAC protocol. Wireless Netw. **7**(2), 159–171 (2001)
26. A. Kumar, E. Altman, D. Miorandi, M. Goyal, New insights from a fixed-point analysis of single cell IEEE 802.11 WLANs. IEEE/ACM Trans. Netw. **15**(3), 588–601 (2007)
27. S.P. Boyd, L. Vandenberghe, *Convex Optimization* (Cambridge University Press, Cambridge, 2004)
28. M. Carvalho, J. Garcia-Luna-Aceves, Delay analysis of IEEE 802.11 in single-hop networks, in *Proc. IEEE ICNP'03*, 2003, pp. 146–155
29. H. Omar, W. Zhuang, A. Abdrabou, L. Li, Performance evaluation of VeMAC supporting safety applications in vehicular networks. IEEE Trans. Emerging Topics Comput. **1**(1), 69–83 (2013)
30. D.P. Bertsekas, R.G. Gallager, P. Humblet, *Data Networks*, vol. 2 (Prentice-hall, Englewood Cliffs, 1987)
31. D. Zheng, Y.-D. Yao, Throughput performance evaluation of two-tier TDMA for sensor networks, in *Proc. IEEE SARNOFF '09*, 2009, pp. 1–5
32. OMNeT++ 5.0. http://www.omnetpp.org/omnetpp
33. Supplement to IEEE Standard for Information Technology Telecommunications and Information Exchange Between Systems Local and Metropolitan Area Networks Specific Requirements Part 11: Wireless LAN Medium Access Control (MAC) and Physical Layer (PHY) Specifications: Higher-Speed Physical Layer Extension in the 2.4 GHz Band, *IEEE Std 802.11b-1999*, pp. i–90, 2000
34. H. Omar, W. Zhuang, L. Li, VeMAC: A TDMA-based MAC protocol for reliable broadcast in VANETs. IEEE Trans. Mobile Comput. **12**(9), 1724–1736 (2013)
35. A. Aijaz, A. Aghvami, Cognitive machine-to-machine communications for Internet-of-Things: a protocol stack perspective. IEEE Internet Things J. **2**(2), 103–112 (2015)
36. G. Bianchi, Performance analysis of the IEEE 802.11 distributed coordination function. IEEE J. Sel. Areas Commun. **18**(3), 535–547 (2000)
37. R. Zhang, L. Cai, J. Pan, Performance analysis of reservation and contention-based hybrid MAC for wireless networks, in *Proc. IEEE ICC'10*, 2010, pp. 1–5
38. R. Zhang, L. Cai, J. Pan, Performance study of hybrid MAC using soft reservation for wireless networks, in *Proc. IEEE ICC'11*, 2011, pp. 1–5
39. P. Wang, H. Jiang, W. Zhuang, Capacity improvement and analysis for voice/data traffic over WLANs. IEEE Trans. Wirel. Commun. **6**(4), 1530–1541 (2007)
40. X. Ling, K.-H. Liu, Y. Cheng, X. Shen, J.W. Mark, A novel performance model for distributed prioritized MAC protocols, in *Proc. IEEE GLOBECOM'07*, 2007, pp. 4692–4696
41. F. Cali, M. Conti, E. Gregori, Dynamic tuning of the IEEE 802.11 protocol to achieve a theoretical throughput limit. IEEE/ACM Trans. Netw. **8**(6), 785–799 (2000)
42. Q. Ye, W. Zhuang, L. Li, P. Vigneron, Traffic load adaptive medium access control for fully-connected mobile ad hoc networks. IEEE Trans. Veh. Technol. **65**(11), 9358–9371 (2016)
43. J. Ren, Y. Zhang, K. Zhang, A. Liu, J. Chen, X. Shen, Lifetime and energy hole evolution analysis in data-gathering wireless sensor networks. IEEE Trans. Ind. Inform. **12**(2), 788–800 (2015)

Chapter 6
Conclusion and Future Works

In this book, we have presented intelligent resource management solutions, from different protocol stack layer perspectives, to improve the efficiency of network slicing for both 5G core and wireless networks. We first give an introduction to the 5G networking architecture for supporting differentiated services in the IoT era, and then describe the concept of network slicing as a new resource orchestration framework, including virtual network topology design, E2E delay modeling for embedded SFCs, transmission and computing resource slicing, and slice-level transport-layer protocol design, to realize fine-grained and customized QoS satisfaction. In particular, for 5G core networks, virtual networks providing specific service/network functionalities are created, in terms of traffic routing topology configuration and virtual function placement, and are embedded over a physical substrate to provide customized service deliveries. To achieve delay-aware virtual network customization, E2E delay is analyzed for each embedded virtual network with properly allocated link transmission and function processing resources. The SDN/NFV-based transport-layer protocols are designed and customized for different sliced networks; For infrastructure-based 5G wireless networks, a dynamic radio resource slicing framework is developed to support a heterogeneity of machine-type and mobile broadband services; For Infrastructure-less wireless networks, we focus on adaptive and service-oriented MAC protocols for packet transmissions in a distributed IoT environment with the ad hoc networking mode.

For future research directions, we discuss in Sects. 6.1 and 6.2 how to use machine learning-based techniques for multi-dimensional resource slicing and protocol generation in a beyond 5G network environment.

Q. Ye, W. Zhuang, *Intelligent Resource Management for Network Slicing in 5G and Beyond*, Wireless Networks, https://doi.org/10.1007/978-3-030-88666-0_6

6.1 Learning-Based Resource Slicing for Beyond 5G Networks

Future communication networks beyond 5G are envisioned to be more complex and highly dynamic, due to largely increased network scales, diversified service and device types, and differentiated mobility patterns. The newly emerged applications, such as autonomous driving, industrial automation, VR/AR, and e-healthcare, often require not only enhanced wireless transmission efficiency but large amount of computing resources (e.g., CPU cores) and caching resources for task processing (e.g., object detection, data classification, data fusion, and fault diagnosis) [1–4]. With the emerging edge intelligence (e.g., edge-cloud computing interplay), it is expected to explore advanced networking and computing solutions to jointly optimize the slicing of multi-dimensional resources (i.e., communication, computing, and caching) to support diversified IoT services for beyond 5G networks [5]. However, the increased network diversity and dynamics pose technical challenges over the conventional model-based resource slicing methodology: (1) The future wireless networks will include not only densely deployed macro-cells and small-cells, but also other inter-networking scenarios such as air-ground integrated networks [6]. With such network complexity, the complete network information is difficult to access, resulting in the challenge for precisely modeling the interaction (e.g., interference) among heterogeneous networks; (2) Conventional resource slicing problems are often formulated as one-shot optimization programs based on instantaneous knowledge of the network information, which is difficult to provide long-term differentiated QoS guarantee under temporal and spatial traffic dynamics [7]; (3) With the advent of edge computing, the resource slicing is required to maximize computing and caching resource utilization [8], which introduces non-convex constraints in different resource optimization frameworks, and thus makes the optimal performance difficult to achieve in real time. Therefore, based on partially-known network information, we aim at investigating deep reinforcement-learning (DRL)-based multi-resource slicing problems to maximize the overall resource utilization with respect to the changing network states and traffic/service demands [2, 9], while guaranteeing differentiated and user-oriented service quality. To handle complex learning tasks, deep neural networks (DNNs) [10] can be investigated to train the optimal resource slicing decisions according to the network environment observation. Timely online solutions can then be obtained based on the well established learning framework.

6.2 Protocol Automation for Beyond 5G Networks

With allocated network resources, it is necessary to investigate how end users/devices access the wireless channels to transmit data packets with high transmission efficiency and satisfied service quality. Therefore, efficient MAC protocols are required to coordinate device transmission behaviors in a future network. Conventional MAC protocols are designed separately for distributed ad hoc networking or for centralized infrastructure-based networking. In a distributed network scenario, each device needs to interact with their neighbors to coordinate transmission behaviors to avoid or resolve transmission collisions; while in a centralized network scenario, a BS or an access point (AP) polls each device for the transmission opportunity allocation. However, with the increased number of end devices and service types (especially IoT services), conventional MAC protocols may not work efficiently, due to high signaling overhead and protocol complexity in making packet transmission decisions. In addition, without complete network information, the protocol performance can degrade, which leads to increasingly accumulated transmission collisions and sub-optimal transmission slot scheduling, and the performance can be further impaired substantially in a dynamic network environment. To achieve near-optimal protocol operation with simplified signaling overhead, one approach is to explore reinforcement learning (RL)-based MAC and transport protocol design, in which the optimal transmission behaviors are iteratively obtained with the set of converged protocol parameters [11, 12]. Every device equipped with the learning automation module makes transmission decisions based on the network state in current time slot, and adjusts the actions in each subsequent time slot based on the response from the network environment (e.g., QoS performance). The RL-based algorithms simplify the procedures (complexity) for protocol generation with partial network state information as the learning inputs, which realize the protocol automation and protocol adaptation to the dynamic networking environment with low signaling overhead.

References

1. Q. Ye, W. Shi, K. Qu, H. He, W. Zhuang, X. Shen, Joint RAN slicing and computation offloading for autonomous vehicular networks: a learning-assisted hierarchical approach. IEEE Open J. Veh. Technol. **2**, 272–288 (2021)
2. Q. Ye, W. Shi, K. Qu, H. He, W. Zhuang, X. Shen, Learning-based computing task offloading for autonomous driving: a load balancing perspective, in *Proc. ICC' 21*, 2021, pp. 1–6
3. W. Wu, P. Yang, W. Zhang, C. Zhou, X. Shen, Accuracy-guaranteed collaborative DNN inference in industrial IoT via deep reinforcement learning. IEEE Trans. Ind. Inform. **17**(7), 4988–4998 (2021)
4. X. Shen, J. Gao, W. Wu, K. Lyu, M. Li, W. Zhuang, X. Li, J. Rao, AI-assisted network-slicing based next-generation wireless networks. IEEE Open J. Veh. Technol. **1**, 45–66 (2020)
5. W. Zhuang, Q. Ye, F. Lyu, N. Cheng, J. Ren, SDN/NFV-empowered future IoV with enhanced communication, computing, and caching. Proc. IEEE **108**(2), 274–291 (2020)
6. H. Wu, J. Chen, C. Zhou, W. Shi, N. Cheng, W. Xu, W. Zhuang, X. Shen, Resource management in space-air-ground integrated vehicular networks: SDN control and AI algorithm design. IEEE Wirel. Commun. **27**(6), 52–60 (2020)
7. M. Zeng, T. Lin, M. Chen, H. Yan, J. Huang, J. Wu, Y. Li, Temporal-spatial mobile application usage understanding and popularity prediction for edge caching. IEEE Wirel. Commun. **25**(3), 36–42 (2018)

8. S. Andreev, O. Galinina, A. Pyattaev, J. Hosek, P. Masek, H. Yanikomeroglu, Y. Koucheryavy, Exploring synergy between communications, caching, and computing in 5G-grade deployments. IEEE Commun. Mag. **54**(8), 60–69 (2016)
9. J. Chen, P. Yang, Q. Ye, W. Zhuang, X. Shen, X. Li, Learning-based proactive resource allocation for delay-sensitive packet transmission. IEEE Trans. Cogn. Commun. Netw. **7**(2), 675–688 (2021)
10. E. Li, L. Zeng, Z. Zhou, X. Chen, Edge AI: on-demand accelerating deep neural network inference via edge computing. IEEE Trans. Wirel. Commun. **19**(1), 4470–457 (2019)
11. P. Nicopolitidis, G.I. Papadimitriou, A.S. Pomportsis, P. Sarigiannidis, M.S. Obaidat, Adaptive wireless networks using learning automata. IEEE Wirel. Commun. **18**(2), 75–81 (2011)
12. R.S. Sutton, A.G. Barto et al., *Reinforcement Learning: An Introduction* (MIT Press, Cambridge, 1998)

Appendices

Appendix A: Proof of Proposition 1

Proof Define $Z^{(k)}$ as packet inter-departure time at the k th server, and define $\rho^{(1)}$ as the probability that the l th departing packet sees a nonempty queue at the first server. If the l th departing packet sees an empty queue at the first server, we let $X^{(1)}$ be the duration from time 0 till the instant that $(l + 1)$ th packet arrives at the server. With $Y^{(q)} \geq Y^{(q-1)}$, there is no queueing delay at each of the servers following the first server. Similar to the description in Fig. 2.13, if the l th departing packet sees a nonempty queue, we have $Z^{(k)} = Y^{(1)}$; Otherwise, two cases are considered:

Case 1   If $X^{(1)} > \sum\limits_{q=2}^{k} Y^{(q)}$,

$$Z^{(k)} = \left(X^{(1)} - \sum_{q=2}^{k} Y^{(q)} \right) + \sum_{q=1}^{k} Y^{(q)} = X^{(1)} + Y^{(1)}; \qquad \text{(A.1)}$$

Case 2   If $X^{(1)} \leq \sum\limits_{q=2}^{k} Y^{(q)}$,

$$Z^{(k)} = \sum_{q=1}^{k} Y^{(q)} - \left(\sum_{q=2}^{k} Y^{(q)} - X^{(1)} \right) = Y^{(1)} + X^{(1)}. \qquad \text{(A.2)}$$

Hence, $Z^{(k)}$ has the same PDF as Y_i derived in (2.34), which ends the proof.

© The Author(s), under exclusive license to Springer Nature Switzerland AG 2021
Q. Ye, W. Zhuang, *Intelligent Resource Management for Network Slicing in 5G and Beyond*, Wireless Networks, https://doi.org/10.1007/978-3-030-88666-0

Appendix B: Proof of Proposition 2

Proof When λ_i is small, σ_i^2 in (2.43) is close to that for an M/D/1 queueing system, given by Bertsekas et al. [1]

$$\sigma_i^2 \approx \frac{1}{\lambda_i^2} - \frac{1}{\mu_{i,1}'^2}. \tag{B.1}$$

Hence, $W_{i,2}$ is further derived as

$$W_{i,2} \approx \frac{\lambda_i \left[\frac{1}{\lambda_i^2} - \frac{1}{\mu_{i,1}^2} - \left(\frac{1}{\lambda_i^2} - \frac{1}{\mu_{i,1}'^2} \right) \right]}{2 \left(1 - \varrho_{i,1}' \right)} = \frac{\lambda_i \left(\frac{1}{\mu_{i,1}'^2} - \frac{1}{\mu_{i,1}^2} \right)}{2 \left(1 - \varrho_{i,1}' \right)}. \tag{B.2}$$

When λ_i becomes large, the idle duration within inter-arrival time of successive packets of flow i at N_2 is small, making σ_i^2 negligible. Thus, we have

$$W_{i,2} \approx \frac{\lambda_i \left(\frac{1}{\lambda_i^2} - \frac{1}{\mu_{i,1}^2} \right)}{2 \left(1 - \varrho_{i,1}' \right)} \approx \frac{\lambda_i \left(\frac{1}{\mu_{i,1}'^2} - \frac{1}{\mu_{i,1}^2} \right)}{2 \left(1 - \varrho_{i,1}' \right)}. \tag{B.3}$$

On the other hand, under both traffic load cases, $Q_{i,2}$ in the approximated M/D/1 queueing system is derived as

$$Q_{i,2} = \frac{\lambda_i}{2\mu_{i,1}'^2 \left(1 - \rho_{i,1}' \right)} \geq W_{i,2}. \tag{B.4}$$

Thus, we prove that $Q_{i,2}$ is an upper bound of $W_{i,2}$ in both lightly-loaded and heavily-loaded traffic conditions, and becomes a tighter upper bound than that in the G/D/1 system in (2.43) when λ_i is small.

Appendix C: Proof of Proposition 3

Proof For brevity, only the proof for (3.13) in Proposition 3 is provided. Since the radio resources, W_s, is reused among all SBSs and the fraction of resources allocated to device i from one SBS is independent of the fraction allocated to device q from another SBS, (S2P1′) can be decoupled into n subproblems, each for one SBS. The subproblem for the SBS B_k ($k \in \{1, 2, \ldots, n\}$) is formulated as

$$(\text{S2P1}' - 1) : \max_{f_{i,k}} u_k^{(1)}(f_{i,k})$$

$$\text{s.t.} \begin{cases} \sum_{i \in \overline{\mathcal{N}_k'}} f_{i,k} = 1 & \text{(C.1a)} \\[2mm] f_{i,k} \in (0, 1), \quad i \in \overline{\mathcal{N}_k'}. & \text{(C.1b)} \end{cases}$$

The objective function of (S2P1′ − 1) can be further derived as

$$u_k^{(1)}(f_{i,k}) = \sum_{i \in \overline{\mathcal{N}_k'}} \log \left(W_v \beta_s f_{i,k} r_{i,k} \right)$$

$$= \log \left(\prod_{i \in \overline{\mathcal{N}_k'}} W_v \beta_s r_{i,k} \right) + \log \left(\prod_{i \in \overline{\mathcal{N}_k'}} f_{i,k} \right). \tag{C.2}$$

Since $r_{i,k}$ is considered constant during each bandwidth slicing period and is independent of $f_{i,k}$, (S2P1′ − 1) is equivalent to

$$(\text{S2P1}' - 2) : \max_{f_{i,k}} \prod_{i \in \overline{\mathcal{N}_k'}} f_{i,k}$$

$$\text{s.t.} \begin{cases} \sum_{i \in \overline{\mathcal{N}_k'}} f_{i,k} = 1 & \text{(C.3a)} \\[2mm] f_{i,k} \in (0, 1), \quad i \in \overline{\mathcal{N}_k'}. & \text{(C.3b)} \end{cases}$$

Since geometric average is no greater than arithmetic average, we have

$$\prod_{i \in \overline{\mathcal{N}_k'}} f_{i,k} \le \left(\frac{\sum\limits_{i \in \overline{\mathcal{N}_k'}} f_{i,k}}{|\overline{\mathcal{N}_k'}|} \right)^{|\overline{\mathcal{N}_k'}|} \tag{C.4}$$

where the equal sign holds when $f_{i,k} = f_{l,k}, \forall i, l \in \mathcal{N}'_k$, and $|\cdot|$ denotes a set cardinality. Thus, by satisfying constraints (C.3a) and (C.3b), the optimal fraction of bandwidth resources allocated to device i associated with S_k ($k \in \{1, 2, \ldots, n\}$) is

$$f^*_{i,k} = \frac{1}{|\mathcal{N}'_k|} = \frac{1}{\sum\limits_{l \in N_k \cup M_k} x_{l,k}} \triangleq f^*_k. \tag{C.5}$$

Similar proof for (3.12) in Proposition 3 can also be made, which is omitted here.

Appendix D: Proof of Proposition 4

Proof Given β_m, we first calculate the Hessian matrix of $u_m^{(2)}(\beta_m, \widetilde{\mathbf{X}}_m)$ with respect to $\widetilde{\mathbf{X}}_m$. That is,

$$\mathbf{H}\left[u_m^{(2)}\left(\beta_m, \widetilde{\mathbf{X}}_m\right)\right] = \begin{bmatrix} -\frac{1}{h(\widetilde{\mathbf{X}}_m)} & -\frac{1}{h(\widetilde{\mathbf{X}}_m)} & \cdots & -\frac{1}{h(\widetilde{\mathbf{X}}_m)} \\ -\frac{1}{h(\widetilde{\mathbf{X}}_m)} & -\frac{1}{h(\widetilde{\mathbf{X}}_m)} & \cdots & -\frac{1}{h(\widetilde{\mathbf{X}}_m)} \\ \vdots & \vdots & \ddots & \vdots \\ -\frac{1}{h(\widetilde{\mathbf{X}}_m)} & -\frac{1}{h(\widetilde{\mathbf{X}}_m)} & \cdots & -\frac{1}{h(\widetilde{\mathbf{X}}_m)} \end{bmatrix} \tag{D.1}$$

where $h(\widetilde{\mathbf{X}}_m) = N_u + N_a + \sum\limits_{k=1}^{n} \sum\limits_{i=1}^{N_k+M_k} \widetilde{x_{i,m}}$, and the dimension of the matrix is $\left[\sum\limits_{k=1}^{n} (N_k + M_k)\right] \times \left[\sum\limits_{k=1}^{n} (N_k + M_k)\right]$.

For any non-zero vector $v = (v_1, v_2, \ldots, v_y) \in \mathbf{R}^y$, $y = \sum\limits_{k=1}^{n} (N_k + M_k)$, we have

$$v^T \mathbf{H}\left[u_m^{(2)}\left(\beta_m, \widetilde{\mathbf{X}}_m\right)\right] v = -\frac{\left(\sum\limits_{i=1}^{y} v_i\right)^2}{h(\widetilde{\mathbf{X}}_m)} < 0. \tag{D.2}$$

Since the Hessian matrix $\mathbf{H}\left[u_m^{(2)}\left(\beta_m, \widetilde{\mathbf{X}}_m\right)\right]$ is negative definite, $u_m^{(2)}(\beta_m, \widetilde{\mathbf{X}}_m)$ is a (strictly) concave function in terms of $\widetilde{\mathbf{X}}_m$ for any fixed β_m. Conversely, it is obvious that $u_m^{(2)}(\beta_m, \widetilde{\mathbf{X}}_m)$ is a (strictly) concave function with respect to β_m for any given $\widetilde{\mathbf{X}}_m$.

Similarly, $u_k^{(2)}\left(\beta_s, \widetilde{\mathbf{X}}_k\right)$ can also be proved as a (strictly) biconcave function. The summation $\sum_{k=1}^{n} u_k^{(2)}\left(\beta_s, \widetilde{\mathbf{X}}_k\right)$ is a nonnegative linear combination of a set of biconcave functions, which is also (strictly) biconcave [2].

Appendix E: Proof of Proposition 5

Proof Property (1) in Proposition 3 can be easily verified for (P3′). To verify the uniqueness of the set of optimal solutions, $\{\beta_m^{(t+1)}, \beta_s^{(t+1)}, \widetilde{\mathbf{X}}_m^{(t+1)}, \widetilde{\mathbf{X}}_k^{(t+1)}\}$, at the end of tth iteration, we refer to the proof of Proposition 2 that, given $\{\beta_m^{(t)}, \beta_s^{(t)}\}$, the objective function of (P3′) is a (strictly) concave function in terms of $\{\mathbf{X}_m, \mathbf{X}_k\}$ and, given $\{\widetilde{\mathbf{X}}_m^{(t+1)}, \widetilde{\mathbf{X}}_k^{(t+1)}\}$, the objective function of (P3′) is also a (strictly) concave function with respect to $\{\beta_m, \beta_s\}$.

Appendix F: Packet Loss Detection Thresholds

We provide details of how packet loss is detected based on the threshold design. Specifically, a packet loss detection depends on the measurements of the time intervals for consecutive packet receptions (denoted by *InterTime*) and the number of received disordered packets (denoted by *InterCnt*), while retransmission delay is measured for retransmitted packet loss detection. If at least one of these measurements exceeds corresponding threshold, a packet loss is detected.

Interarrival Timeout The time interval between consecutive packet receptions is interarrival time, which indicates the conditions of link congestion and packet loss. At a retransmission node, there is an elapsed timer initiating at the instant of receiving a packet. If time duration is longer than a threshold, *expected interarrival time*, a packet loss is detected due to interarrival timeout. We obtain the expected interarrival time by linear prediction, based on sampled interarrival time durations at the retransmission node.

Interarrival Counter Threshold Since the packets from one service slice are forwarded following the linear topology, out-of-order packet reception at a retransmission node indicates packet loss. After a packet loss is detected by a retransmission node, an RR packet is sent from the retransmission node to its upstream caching nodes, and an RD packet will be generated and sent by a caching node. When the RD packet is received by the downstream retransmission nodes, it leads to out-of-order packet reception. Thus, a retransmission node can detect packet loss depending on the level of packet disorder (i.e., packet disorder length).

To avoid spurious packet loss detection, the retransmission nodes should estimate an updated packet disorder length as threshold for InterCnt in the expected list. To do so, each retransmission node needs to determine where a packet loss actually happens. If the packet loss happens in its own segment, the retransmission node sends an RR packet to trigger retransmission; Otherwise, the retransmission node estimates an updated disorder length for a retransmitted packet to avoid duplicate retransmissions. The updated disorder length is set as the interarrival counter threshold (CntThres) for the packet in the expected packet lists.

Based on signaling exchange with other retransmission nodes, one retransmission node can update the packet-level thresholds that are differentiated for each packet. To reduce the signaling overhead, the retransmission node can also obtain the segment-level threshold through sampling and estimation, based on the InterCnt of received out-of-order packets.

Retransmission Timeout After packet loss is detected, RR and RD packets are transmitted for loss recovery. Since packet loss can happen to both RR and RD packets, the node that triggers a retransmission should be able to detect the retransmitted packet loss and resend an RR packet. To detect the loss, we use the expected retransmission delay as a threshold for RTTimer in the expected list, in which the estimation is based on the sampled packet retransmission delay. The time of sending an RR packet is recorded in its timestamp field, and is included in the corresponding RD packet when it is generated. When the RD packet is received at a retransmission node, the retransmission delay is calculated, i.e., the duration from the time instant the RR packet is sent till the instant the RD packet is received. The calculation of thresholds, including the expected interarrival time, packet-level and segment-level CntThres, and the expected retransmission delay, is presented in detail in our previous work [3].

Appendix G: Methodology for Packet Loss Detection

After the connection establishment, a *content window list* is established and maintained at each retransmission node, to record the packets successfully received at the node. At the same time, an *expected packet list* is established to record the packets that are expected to be received at the retransmission node. If packet loss is detected, contents in the expected packet list are referred for triggering packet retransmissions.

Content Window List In the content window list, packets received in sequence are described by one window. Each window is bounded by its *left edge* and *right edge*, which indicate the first sequence number and the next expected sequence number of the received packets respectively. Thus, the sequence numbers of sequentially received packets lie between left edge and right edge of one content window. Due to packet loss, the packets may not be received in sequence, and a new window is generated in the list. Whenever a new packet is received, the content window list is updated by the retransmission node.

Expected Packet List To record the information of packets that are expected by a retransmission node, the expected packet list has following fields:

1. *Num*—When packets are lost discontinuously, expected packets are in separated packet windows, which are described by different entries in the expected list. The Num field represents the sequence of the entries;
2. *StartSeq* and *EndSeq*—Specify an sequence number interval with a start and an end sequence numbers;
3. *StartNum*—Indicate the number of packet offset between the target packet and the packet with StartSeq as its sequence number;
4. *InterCnt*—Record how many packets have been received after the last sequentially received packet, when packets are received discontinuously;
5. *CntThres*—Indicate the threshold set for InterCnt in packet loss detection;
6. *WaitLen*—Measure the difference between CntThres and InterCnt;
7. *RTCnt*—Record the number of retransmission requests sent for lost packets;
8. *RTType*—Indicate how packet loss is detected by packet interarrival time exceeding a timeout, InterCnt greater than CntThres, or the duration from triggering the retransmission request to receiving the retransmitted packet exceeding a timeout;
9. *RTTimer*—Denote an elapsed timer starting from the time instant of sending a retransmission request. If this time duration is larger than a retransmission timeout, another retransmission request of the lost packet is triggered.

The expected packet list is established based on the corresponding content window list at a retransmission node. For each entry in the expected packet list, StartSeq is set as the value in the right edge field of corresponding content window, and EndSeq is set as the maximum sequence number less than the left edge of its subsequent content window. In the case of a determined EndSeq, the packets with the sequence numbers lying between StartSeq and EndSeq are expected; In the case of an undetermined EndSeq (set as infinity), the expected packet is located by using StartSeq and StartNum. As the end sequence number for the last content window cannot be determined, we set StartNum as 0 to indicate the expected packet after the last content window. When a new entry is established, we initialize CntThres and WaitLen to 1, and the other fileds to 0.

When a caching node receives an RR packet from a retransmission node, it generates an RD packet whose header includes the StartSeq and StartNum information from the received RR packet. When receiving the RD packet, the retransmission node not only uses the RD sequence number, but also uses the StartSeq and StartNum pair to match the entries in the expected packet list. After an entry in the list is matched, it is removed from the list, indicating that the RD packet is successfully received. If the packet interarrival time at a retransmission node is longer than a threshold, indicating consecutive packet loss, the expected packets after the last entry of the content window list will be requested for retransmission one by one. StartSeq and StartNum in the expected packet list are used to locate each expected packet. Whenever packet loss is detected due to packet interarrival timeout, a new entry is added to the expected packet list, where StartSeq remains unchanged, and StartNum is incremented by one (pointing to the subsequent expected packet).

The header format of an SDATP packet is shown in Fig. 4.2, in which the Flag field indicates whether an RR packet is triggered by interarrival timeout, exceeding an interarrival counter threshold, or is triggered by retransmission timeout. For packet loss detected by exceeding interarrival counter threshold and retransmission timeout, both StartSeq and EndSeq are specified in the RR packet, indicating the range of sequence numbers of lost packets. For packet loss detected by interarrival timeout where EndSeq is unknown, StartSeq and StartNum in the RR packet are used to locate a specific expected packet for retransmission. The time instant of generating this RR packet is represented by the Timestamp field.

Data traffic from different slices may go though a common network switch sharing a set of resources. However, each slice has an independent set of content window list, expected packet list, and the associated parameters and variables maintained at each retransmission node.

Appendix H: Proof of Proposition 6

Proof The objective function of the logarithmic nonlinear least-squares curve-fitting problem can be written as

$$||a_1 + a_2 \ln(\mathbf{N}) - \mathbf{P}||_2^2 = \sum_{n=2}^{N} (a_1 + a_2 \ln(n) - p_n)^2 = \sum_{n=2}^{N} f_n^2(\mathbf{a}). \quad \text{(H.1)}$$

Then, $\forall \mathbf{a} \in \mathbf{dom}\, f_n$, we calculate the Hessian matrix of $f_n^2(\mathbf{a})$ as follows:

$$\mathbf{H}(f_n^2(\mathbf{a})) = \begin{bmatrix} \frac{\partial f_n^2(\mathbf{a})}{\partial a_1^2} & \frac{\partial f_n^2(\mathbf{a})}{\partial a_1 \partial a_2} \\ \frac{\partial f_n^2(\mathbf{a})}{\partial a_2 \partial a_1} & \frac{\partial f_n^2(\mathbf{a})}{\partial a_2^2} \end{bmatrix} = \begin{bmatrix} 2 & 2\ln(n) \\ 2\ln(n) & 2\ln^2(n) \end{bmatrix}. \quad \text{(H.2)}$$

The eigenvalues of the Hessian matrix can be derived by solving the eigenfunction of $\mathbf{H}(f_n^2)$

$$\det(\lambda\,\mathbf{I} - \mathbf{H}(f_n^2)) = \begin{vmatrix} \lambda - 2 & -2\ln(n) \\ -2\ln(n) & \lambda - 2\ln^2(n) \end{vmatrix} = 0 \quad \text{(H.3)}$$

$$\implies \lambda_1 = 0, \lambda_2 = 2 + 2\ln^2(n).$$

Because both eigenvalues of $\mathbf{H}(f_n^2)$ are nonnegative, the Hessian matrix $\mathbf{H}(f_n^2)$ is semidefinite. On the other hand, since $\mathbf{dom}\, f_n = \{(a_1, a_2) \mid a_2 \geq 0\}$ is a convex set, $f_n^2(\mathbf{a})$ is a convex function for all $\mathbf{a} \in \mathbf{dom}\, f_n$.

Hence, the objective function $\sum_{n=2}^{N} f_n^2(\mathbf{a})$ is a nonnegative sum of convex functions $f_n^2(\mathbf{a})$ $(n = 2, 3, \ldots, N)$, which is also convex [4]. That is, the curve-fitting is a convex optimization problem.

Appendix I: Derivation of $E[W_{st}]$ and $E[W_{qt}]$

In order to calculate the average packet delay of the M/G/1 queue of a tagged node, we first derive the probability distribution of packet service time W_{st}. For analysis simplicity, we normalize the control period of each frame as an integer multiple of one D-TDMA data slot duration T_p, i.e., $M_c = \lceil \frac{M_m T_m}{T_p} \rceil$, where $\lceil \rceil$ is the ceiling function. The end instant of each slot along one D-TDMA frame is numbered from 1 to $M_c + N$, as shown in Fig. 1. Let random variable J denote the arriving instant of each head-of-line (HOL) packet. It is assumed that HOL packets of the tagged node only appear at the end of each time slot, neglecting the possibility that HOL packets can arrive within the duration of each time slot [5], which means J takes discrete values from set $\mathbf{A} = \{1, 2, \ldots, M_c + N\}$. From Fig. 1, it can be seen that HOL packets have two different arriving patterns according to current status of the queue: (a) when the node's queue is non-empty (i.e., at least one packet staying in the queueing system), HOL packets can only appear at the end of its designated data slot in the data transmission period, which means J takes values from set $\mathbf{A}' = \{M_c+1, M_c+2, \ldots, M_c+N\}$; (b) when the node's queue is empty (i.e., no packets are in service.), HOL packets can arrive at any time instant in set $\mathbf{A}$. Next, we derive the distribution of W_{st} under these two cases.

When the node's queue is non-empty, based on the assumption that data slot is randomly selected for each node in the next frame upon the successful packet

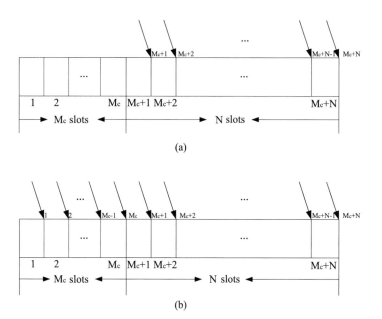

(a)

(b)

Fig. 1 The HOL packets arrival patterns within one frame. (**a**) The node's queue is non-empty. (**b**) The node's queue is empty

transmission in the current frame, we use random variable I, which takes values from set $\mathbf{B} = \{1, 2, \ldots, N\}$, to denote the data slot number that the node selects in the next frame. Thus, the probability distribution of the packet service time W_{st} in the unit of one data slot duration, denoted by W_s, is derived as

$$P\{W_s = M_c + k\} = \sum_{j \in A', i \in \mathbf{B}} P\{W_s = M_c + k, J = j, I = i\}$$

$$= \sum_{j \in A', i \in \mathbf{B}} P\{W_s = M_c + k | J = j, I = i\} P\{J = j\} P\{I = i\} \tag{I.1}$$

$$= \frac{k}{N^2} \quad (1 \le k \le N);$$

$$P\{W_s = M_c + N + k\} = \frac{N - k}{N^2} \quad (1 \le k \le N - 1).$$

When the node's queue is empty, HOL packets can arrive at any time instant in set $\mathbf{A}$. Thus, the probability distribution of W_s is derived in the following two cases:

(i) If $M_c \ge N$,

$$P\{W_s = k\} = \sum_{j=M_c-k+1}^{M_c-k+N} P\{W_s = k | J = j\} P\{J = j\} = \frac{1}{M_c + N}$$

$$(1 \le k \le N - 1)$$

$$P\{W_s = M_c - k\} = \sum_{j=k+1}^{k+N} P\{W_s = M_c - k | J = j\} P\{J = j\} = \frac{1}{M_c + N}$$

$$(0 \le k \le M_c - N)$$

$$P\{W_s = M_c + k\} = \sum_{j=1}^{N-k} \sum_{j=M_c+N-k+1}^{M_c+N} P\{W_s = M_c + k | J = j\} P\{J = j\} = \frac{1}{M_c + N}$$

$$(1 \le k \le N);$$

$$\tag{I.2}$$

(ii) If $M < N$,

$$P\{W_s = k\} = \sum_{j=M_c-k+1}^{M_c-k+N} P\{W_s = k | J = j\} P\{J = j\} = \frac{1}{M_c + N}$$

$$(1 \leq k \leq M_c)$$

$$P\{W_s = M_c + k\} = \sum_{j=1}^{N-k} \sum_{j=M_c+N-k+1}^{M_c+N} P\{W_s = M_c + k | J = j\} P\{J = j\} = \frac{1}{M_c + N}$$

$$(1 \leq k \leq N).$$

(I.3)

Hence, the average service time, $E[W_{st}]$, and the second moment of service time, $E[W_{st}^2]$, are derived as follows:

$$E[W_{st}] = P_{qn} \cdot \sum_{k_1 \in C} k_1 T_p P\{W_s = k_1\} + P_{qe} \cdot \sum_{k_2 \in D} k_2 T_p P\{W_s = k_2\}$$

$$= \lambda E[W_{st}] \cdot \sum_{k_1 \in C} k_1 T_p P\{W_s = k_1\} + (1 - \lambda E[W_{st}]) \cdot \sum_{k_2 \in D} k_2 T_p P\{W_s = k_2\}$$

$$\Longrightarrow E[W_{st}] = \frac{(M_c + N + 1)T_p}{2 - \lambda(M_c + N - 1)T_p};$$

(I.4)

$$E[W_{st}^2] = \frac{(2M_c + 2N + 1)(M_c + N + 1)T_p^2}{6} + T_p^2 \lambda E[W_{st}] \cdot$$

(I.5)

$$\left[(M_c + N)^2 + \frac{N^2 - 1}{6} - \frac{(2M_c + 2N + 1)(M_c + N + 1)}{6} \right]$$

where P_{qe} is the queue empty probability; $\mathbf{C}$ and $\mathbf{D}$ are two sets of possible values of W_s for the queue non-empty and queue empty cases, respectively.

References

1. D.P. Bertsekas, R.G. Gallager, P. Humblet, *Data Networks*, vol. 2 (Prentice-Hall, Englewood Cliffs, 1987)
2. J. Gorski, F. Pfeuffer, K. Klamroth, Biconvex sets and optimization with biconvex functions: a survey and extensions. Math. Methods Oper. Res. **66**(3), 373–407 (2007)
3. J. Chen, S. Yan, Q. Ye, W. Quan, P.T. Do, W. Zhuang, X. Shen, X. Li, J. Rao, An SDN-based transmission protocol with in-path packet caching and retransmission, in *Proc. IEEE ICC*, 2019, pp. 1–6
4. S.P. Boyd, L. Vandenberghe, *Convex Optimization* (Cambridge University Press, Cambridge, 2004)
5. H. Omar, W. Zhuang, A. Abdrabou, L. Li, Performance evaluation of VeMAC supporting safety applications in vehicular networks. IEEE Trans. Emerging Topics Comput. **1**(1), 69–83 (2013)

Index

© The Author(s), under exclusive license to Springer Nature Switzerland AG 2021
Q. Ye, W. Zhuang, *Intelligent Resource Management for Network Slicing in 5G
and Beyond*, Wireless Networks, https://doi.org/10.1007/978-3-030-88666-0

Printed in the United States
by Baker & Taylor Publisher Services